# A SHORT HISTORY
## OF
# NAVAL AND MARINE
# ENGINEERING

PLATE I

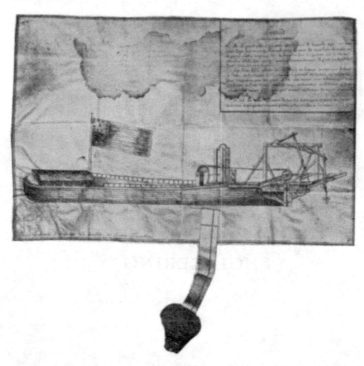

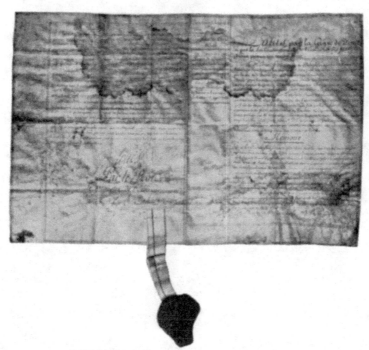

FITCH'S FRENCH PATENT, 1791
Courtesy of United States National Museum

# A SHORT HISTORY
# OF NAVAL AND MARINE
# ENGINEERING

by

ENG. CAPT. EDGAR C. SMITH, O.B.E., R.N.

With a Foreword by
MAJOR P. J. COWAN, M.B.E.

CAMBRIDGE
AT THE UNIVERSITY PRESS
1938

CAMBRIDGE UNIVERSITY PRESS
Cambridge, New York, Melbourne, Madrid, Cape Town,
Singapore, São Paulo, Delhi, Mexico City

Cambridge University Press
The Edinburgh Building, Cambridge CB2 8RU, UK

Published in the United States of America by Cambridge University Press, New York

www.cambridge.org
Information on this title: www.cambridge.org/9781107672932

First published 1938
First paperback edition 2013

*A catalogue record for this publication is available from the British Library*

ISBN 978-1-107-67293-2 Paperback

TO MY
WIFE, SONS and
DAUGHTER

# A SHORT HISTORY OF NAVAL AND MARINE ENGINEERING

## ERRATA

Page 112, line 29, *for* Pl. V *read* Pl. VI

,, 118, ,, 30, *for* Pl. V *read* Pl. VI

,, 181, ,, 29, *for* engines *read* pressure

,, 181, ,, 30, *for* pressure *read* engines

,, 218, ,, 1, *for* 1819 *read* 1810

Pl. IX, facing page 247, *for* ENGINE *read* ENGINES

Page 247, line 13, *for* Pl. X *read* Pl. IX

,, 309, ,, 9, *for* 1891 *read* 1881

,, 321, ,, 25, *for* figure *read* figures

,, 322, ,, 16, *insert* 17,000 *before* horse-power

,, 362, Babcock and Wilcox Boilers, *for* 51 *read* 251

,, 371, Normand, J. A., *delete* 223

# CONTENTS

Coat Button
worn by Engineers
in Royal Navy
1841 to 1856.

# LIST OF PLATES

# LIST OF FIGURES IN THE TEXT

# FOREWORD

SHOULD any commendation of this volume be deemed desirable by possible readers, it affords me great pleasure to testify, from long association with my friend the author, to his qualifications for the task he set himself in its preparation.

Engineer Captain Edgar C. Smith is largely endowed with the traits so necessary to success in such a venture, and during the course of the preparation of the material, most of which, spread over some years, has appeared in the pages of *Engineering*, I can vouch for the careful manner in which clues have been followed up, sources of information traced and the value of "facts" brought to light appraised. Some of the phases dealt with have, of course, been the subject of previous memoirs—often of a somewhat biased nature. It has been Captain Smith's endeavour throughout to weigh evidence fairly and to arrive at a just conclusion. Material has thus been brought into juxtaposition for the first time, with the result that new light is often thrown upon old subjects, while the fairness with which international claims have been handled will be found to be a marked characteristic of the chapters into which these enter.

The actual collection of material for the work was commenced more than thirty years ago, and in the intervening period Captain Smith has been able to track down many elusive points. Starting in one direction, a clue has often put the searcher on to other and in some cases unexpected sources, until the author has been able to feel assured that the pieces of his puzzle fit true and snugly together. It has been of great interest to me to watch this evolution, and that the finished work will be read with interest may be taken as proved by the notice which the original articles in *Engineering* attracted at home and abroad as they appeared. In view of the proximate centenary of the Transatlantic Service, the appearance of the volume is opportune, while it is no less fitting that its contents should be brought together in volume form and in this way be made more readily accessible to the wide circle it deserves as an historical reference book.

P. J. COWAN

*Engineering*
35 & 36 Bedford Street
Strand, London, w.c. 2

# PREFACE

BACON once wrote, "I hold every man a debtor to his profession from the which as men of course do seek to receive countenance and profit so ought they of duty to endeavour themselves to be a help and an ornament thereto." With this saying in mind this book has been written as a contribution to the great branch of engineering in which I have been engaged, and with the hope that it will prove of interest to many of the thousands concerned with ships and their machinery.

Its publication has been rendered possible by the kind permission of the Editor of *Engineering*, Major P. J. Cowan, to include in it the greater part of the series of articles I had the privilege of contributing to that journal during 1930–3 entitled "Chapters in the History of Naval and Marine Engineering". Chapters XX, XXI and XXII, however, appear for the first time. To him I am also indebted for the Foreword.

The book is an attempt to fill a gap in the literature of marine engineering, but no claim is made that it is in any way an exhaustive history of the subject. The material on which it is based has been gathered together during a long period and from many sources. Technical papers, presidential addresses, journals, textbooks, biographies, official regulations, personal letters, reminiscences and hitherto unpublished manuscripts have all been drawn upon to illustrate the many aspects of naval and marine engineering, and their influence on naval warfare and sea transport. Efforts have been made to do justice to the pioneers who took those first steps, which are always the most difficult, remembering the saying of Edward Lear that "if a man does anything at all through life with a deal of pother, and likewise of some benefit to others, the details of such pother and benefit may as well be known accurately as the contrary". As the book is admittedly of a popular character it has not been considered desirable to give a mass of detailed references, but sufficient have been included to assist those who wish to pursue the subject further.

In bringing to a close the task I set myself, it is both my pleasure and my duty to record my sincere thanks to all those who, during the last twenty or thirty years, have assisted me in one way or another. The pleasure is not untinged with sadness, for as I turn over my letters, I meet with signature after signature of those I can no longer thank in person. I have been helped by the loan or gift of books, documents, photographs and drawings, and by first-hand accounts of personal experiences. For example Inspector of Machinery Matthew McIntyre, who died in 1931 at the age of 93, when nearly eighty years of age wrote me an account of his whole career in the Royal Navy, which began in 1859 in a ship with boilers working at 10 lb. pressure and with engines with wooden-toothed gearing. It would be quite impossible to mention all those I should like to, but I may perhaps especially thank Engineer Rear-Admiral Sir Robert Beeman, Mr Charles de Grave Sells, M. Paul Augustin-Normand, Mr W. A. Tookey, Mr A. A. Gomme of the Patent Office Library and Mr H. W. Dickinson; the last named has read all the chapters both in manuscript and in proof and his assistance has been invaluable.

I have also been assisted by various Government departments, technical institutions and societies, and by many firms. For illustrations in the book I have to thank Babcock and Wilcox, Limited; Cammell, Laird and Co., Limited; the Chief Librarian of Birmingham; the Editor of *Engineering*; The Fairfield Shipbuilding and Engineering Co., Limited; Harland and Wolff, Limited; the Maschinenfabrik Augsburg-Nürnberg A.G.; The North-Eastern Marine Engineering Co., Limited; Sautter-Harlé and Co. of Paris; John I. Thornycroft and Co., Limited; Vickers-Armstrongs, Limited; and Yarrow and Co., Limited.

I have also to thank Mr Carl W. Mitman of the United States National Museum for the photograph of Fitch's patent of 1791; the American Society of Mechanical Engineers for the photograph of the bust of Admiral Isherwood; the Deutsches Museum, Munich, for the portrait of Rudolph Diesel; Mr James Napier for permission to reproduce the portrait of his great-grandfather, Robert Napier; Blackie and Son, Limited, for permission to reproduce the portrait of Herbert Akroyd Stuart, and the Director of the National Portrait Gallery for permission to reproduce the portrait of I. K. Brunel. The storehouse of engineering history is, as every student knows, the Science Museum, South Kensington. I have made free use of its publications and to its Director I am indebted for the portraits of Ericsson and Sir F. P. Smith, the illustrations of the *Archimedes*, *Great Western*, *Britannia* and *Rattler* and *Alecto*, and for the cross-sectional sketch of the *Great Eastern*.

EDGAR C. SMITH

PURLEY, SURREY
*July* 1937

## CHAPTER I

# THE BIRTH OF THE STEAM BOAT

AMONG the many branches of engineering which during the last century and a half have widely and profoundly influenced the progress of civilisation, that of marine engineering has an interest second to none. More than a century ago Tredgold, in the elaboration of his memorable definition of civil engineering as "the art of directing the great sources of power in Nature for the use and convenience of man", included "the art of navigation by artificial power for the purposes of commerce", but of this art he lived only long enough to see the first-fruits. When he died in 1829 steam vessels were still small in size and comparatively few in number, and the greater part of the world's overseas commerce was then, and for many years afterwards, carried in ships driven by the wind; within half a century, owing to the advance of marine engineering, the sailing ship was fighting hard for its very existence. In those fifty years the marine steam engine had brought about a revolution in methods of sea transport and in naval warfare, and marine engineering had become a great national industry.

Marine engineering, it is true, is but a branch of mechanical engineering which had its birth in the eighteenth century with the work of Newcomen and Watt, and most of the early marine engines differed little from land engines; but the application of steam power to ships presented so many new problems, gave rise to so many new inventions, engrossed the attention of so many great engineers and led to such splendid achievements that its history claims special consideration. Then, too, it must not be forgotten that the work of the early marine engineers led to a revolution in the design and construction of the ships themselves, and naval architects and shipbuilders were obliged by force of circumstances to adopt some of the methods of the engineers. It was engineering which gave rise to the use of iron for the hulls of ships and to the use of iron armour for fighting vessels; the liners and battleships of to-day are, in the main, the results of the application of engineering to ships which began with the work of the pioneers of a century and a half ago.

Though it would be wrong to neglect what was done in the United States, the progress of marine engineering was mainly due to the work of British engineers. As soon as it had been demonstrated that boats could be successfully driven by steam, establishments for the construction of marine engines sprang up here and there. In the course of time from the Clyde, the Thames, the Mersey and the Tyne, from places like Southampton, Bristol and Hull, went steam ships for practically every nation, until at one time 80 per cent of the world's sea-going steam vessels came from British yards. With but few exceptions, all the greatest improvements in steamship construction and marine engineering were originated in this country, and though other nations founded their own ship yards and engine factories, it was often done with the assistance of British engineers. To-day the ships of America, France, Germany, Italy, Japan and other countries, in size, speed, equipment and performance, rival any we can produce, but it may be presumed that, even if our insular position did not constitute a constant stimulus to fresh activity, the genius of our race for nautical affairs will lead our constructors of the future to emulate the achievements of their predecessors. With the future, however, we are not here concerned—if engineering ever has a deity it should, like Janus, have two faces, one looking forward, one backward. It was no less a man than Pasteur who recalled to the students of Edinburgh the exhortation to "remember the past and look to the future", impressing upon them the advice to "associate the cult of great men and great things with every thought". No branch of engineering is associated more closely with great men and great ideas than that of marine engineering, and it is one of the objects of the following chapters to recall some of these to mind.

The practicability of the application of the steam engine to boats was demonstrated almost simultaneously in America, Scotland and France. In America the principal experiments were made by Rumsey, Fitch, Stevens and Fulton; in Scotland by Miller, Symington and Bell; and in France by Périer and the Marquis de Jouffroy d'Abbans. These trials were all carried out during the last two decades of the eighteenth century and the early years of the nineteenth century. There had been earlier pioneers who had planned and schemed but with no success. There are also stories of a fabulous nature such as that relating to Blasco de Garay, who is said to have moved a boat by steam at Barcelona in 1543. A certain Thomas Gonzales in 1825 was apparently the first to credit Blasco de Garay with using steam, but the documents relating to the incident were examined by John Macgregor who, in a paper read before the Royal Society of Arts on April 14, 1858, stated that "neither of them contained any mention whatever of the use of steam".

About a century after Blasco de Garay's experiments

with boats, David Ramsey obtained a patent in this country which referred to the use of fire and its application to boats, ships and barges. The names of the Marquis of Worcester, Denis Papin and others are sometimes coupled with the invention of the steam boat, but nothing came of their labours. Another projector was Jonathan Hulls, of Campden, Gloucestershire, who on December 21, 1736, secured a patent, and in 1737 published a *Description and Draught of a new-invented Machine for carrying Vessels or Ships...against Wind and Tide or in a Calm*, detailing his invention of the principle of steam navigation. There appears to be no evidence that Hulls actually constructed a steam boat and, as the only steam engine available at that time was the cumbrous slow-moving Newcomen atmospheric engine, it is exceedingly improbable that he could have driven a boat by steam.

The experiments made at the end of the eighteenth century are in a different category from those just referred to, and in most cases are supported by documentary evidence. The accounts of the experiments made in France by Périer and the Marquis de Jouffroy d'Abbans are the least satisfactory, and the experiments cannot be said to have had much influence on the introduction of steam navigation. Jacques Constantin Périer (1742–1818) was a very able engineer, while the Marquis de Jouffroy d'Abbans (1751–1832) was an infantry officer. Between 1770 and 1780 Périer is said to have tried a steam boat on the Seine, while in 1776 and in 1783 the Marquis de Jouffroy d'Abbans made experiments on the River Doubs and the River Saône respectively. By that time the steam engine was attracting much attention and there is no difficulty in believing that the Marquis de Jouffroy d'Abbans obtained some measure of success. Unfortunately no account of his work was published till thirty years later, and then another twenty-five years elapsed before his claims were examined by the Paris Académie des Sciences. In the absence of contemporary documentary evidence it is therefore impossible to say with certainty what machinery he employed or what results he obtained.

But whatever was done in France was far surpassed in America by the work of James Rumsey (1743–92) on the Potomac, and of John Fitch (1743–98) on the Delaware. Biographers have kept the memories of both these inventors alive, and a few years ago Mr L. F. Loree, of the Delaware and Hudson Company, placed students of steam boat lore under an obligation by the reproduction of parts of documents, books and pamphlets relating to their projects, under the title *Developments of Steam Navigation*. The Catalogue of the Water Craft Collection of the National Museum, Washington, contains valuable notes on the work of Fitch and Rumsey, while among the latest reviews of the work of Fitch is that contained in a memoir by M. Paul Augustin-Normand, read on June 16, 1933, to the French Académie de Marine.

Rumsey hailed from Bohemia Manor, Cecil Co., Maryland. The first definite information of his experiments on propulsion is contained in a note in the diary of Washington who, on September 6, 1784, wrote "Remained at Bath all day and was shown the model of a boat constructed by the ingenious Mr Rumsey for ascending rapid currents by mechanism". Evidence goes to show that Rumsey's first plan was to work his boats up-stream by mechanically-worked poles, but this plan was later abandoned for the system proposed long before, and since known as jet propulsion, in which water drawn into a pump in the boat is ejected at the stern. The first trial of any consequence appears to have been made on December 3, 1787, and another trial was made a few days later. In the main Rumsey's machinery consisted of a boiler, a cylinder and a pump, the cylinder and pump barrel being placed vertically, the one on top of the other. One of the boilers tried was a "pipe boiler".

Besides his ideas for steam boats Rumsey put forward projects for pumping, for mills, etc., all appearing to be generally useful, and in 1788 the Rumseian Society was formed to further his schemes, Benjamin Franklin being one of its supporters. In 1788 also, Rumsey published at Philadelphia *A Short Treatise on the Application of Steam whereby is Clearly Shown from Actual Experiments that Steam may be Applied to Propel Boats or Vessels of any Burthen against Rapid Currents with great Velocity*.... The formation of the Rumseian Society led the inventor to visit England. In London he took out two patents, No. 1669 of November 6, 1788, and No. 1734 of March 24, 1790. When applying for the second of these he described himself as of the parish of Saint Margaret, Westminster. In 1792 he tried a steam boat on the Thames, to which Fulton made a reference in his note-book. All further action, however, was brought to an end by Rumsey's unexpected death on December 23, 1792, at the age of forty-nine, in London. He was buried in St Margaret's churchyard. Many years afterwards, in 1839, the Kentucky Legislature presented a gold medal to his son "Commemorative of his father's services and high agency in giving the World the benefit of the steam boat".

Rumsey's great rival John Fitch came from Windsor, Connecticut. It is said he first conceived the idea of applying steam to transport in April 1785. In September of that year he laid a drawing, a description and a model of a steam boat before the American Philosophical Society, Philadelphia. In October he had an interview with Washington and soon afterwards applied to the legislature of Virginia for assistance. In March 1786 he obtained exclusive privileges for his boats in New Jersey, and during the next eighteen months obtained similar privileges from the States of Delaware, New York, Virginia and Pennsylvania. During the three years 1785 to 1788 he built at least three steam boats, 34 ft., 45 ft. and 60 ft. long respectively. A model of the first of these is in the National Museum. In this boat Fitch used twelve vertical oars like snow shovels, operated by gears, and a sprocket wheel and chain driven by a steam engine. In 1788 a Steam Boat Company was formed and his 60 ft. boat made a trip from Philadelphia to Burlington, about 20 miles. After a trip in this boat on October 12, 1788, ten worthies of Philadelphia declared "we are clearly of opinion that the rivers of America may be navigated by the means of steam boats, and that the present boat would be very useful on the Western Waters". Two years later Fitch placed another vessel on the Delaware, which ran for three or four months. An advertisement of July 26, 1790, announced

## THE STEAMBOAT

Sets out tomorrow morning at ten o'clock, from Arch Street Ferry, in order to take passengers for Burlington, Bristol, Bordentown, and Trenton, and return next day.

Fitch and his supporters next planned to exploit his invention in Europe and through Aaron Vail, United States Consul at L'Orient, France, on November 29, 1791, secured a French patent. It is of interest to note that this patent was one of the very early ones granted under the Patent Law which had just been brought into force. The original of Fitch's Letters Patent, signed by Louis XVI and bearing the great seal of France, now in the United States National Museum, Washington, D.C., it is stated, was rescued from the wreckage of the National Library after the Commune and was taken to America. It is shown on Plate I and a translation of the essential part of the patent is given below.

DIRECTORY OF INVENTION No. 28

DEPARTMENT OF PARIS
16  2h.

29 November 1791

Letters patent for Fifteen Years, for a mechanism suitable for making boats, ships and other vessels move by means of a fire engine.

TO MR JOHN FITCH, citizen of Philadelphia, in the United States of America.

INVENTOR
AND IMPORTER.

LOUIS, by the Grace of God, and by the Constitutional Law of the State, King of the French, to all present and to come, Greeting.

MR JOHN FITCH, citizen of Philadelphia in the United States of America, having made known to us that he desires to possess the rights of property assured by the law of January 7, 1791, to promoters of discoveries and inventions in every kind of industry and consequently to obtain a patent of invention which shall last for the period of fifteen years, for the exercise and use in all the Kingdom of a MECHANISM SUITABLE TO MAKE BOATS, SHIPS AND OTHER VESSELS MOVE BY MEANS OF A FIRE ENGINE, of which mechanism the said Mr Fitch is declared to be the INVENTOR and IMPORTER; also that it follows from the report established at the time of the deposition made by Mr Aaron Vail, given powers of attorney by the said John Fitch, to the Secretariat of the Directory of the Department in Paris, where the said powers have been communicated under date of the twenty sixth November seventeen hundred and ninety one. Having seen the request of the said Mr Vail, in the name of Mr Fitch, together with the explanatory memorandum and the drawing, addressed by the petitioners to the DIRECTORY OF PATENTS FOR INVENTIONS of which memorandum and drawing follow the literal text and the copy.

MEMORANDUM

Mr John Fitch, citizen of Philadelphia, in the United States
of America, is the inventor of a machine suitable to make
boats, ships and other vessels move, where fire serves as the
motive power, operating by means of boilers, cylinders,
pistons, air pumps, wheels and iron chains, like the drawing
here annexed and according to any other, either in the application of vapours, smoke or fire to navigation in all operations
whatsoever, in following up which inventions and perfec-
tions the said Mr Fitch has made considerable sacrifice,
expending very large sums, as well as in work lasting six con-
secutive years, for compensation of which Mr Fitch is recognised in the procuration of Mr Aaron Vail, citizen of New
York, now in this city, with a view of soliciting the Government of France to grant to him a patent of invention for
fifteen years comformable to the laws of 7 of January and of
25 May 1791.

Paris November 25, 1791.

(Signed) Aaron Vail

\*   \*   \*   \*   \*   \*   \*
\*   \*   \*   \*   \*   \*   \*

WE MAKE KNOWN to and command all tribunals ad-
ministrative bodies and municipalities that they make John
Fitch and his assigns to enjoy and use fully and peacefully the
rights conferred by these presents ceasing and causing to
cease all troubles and contrary hindrances; we make known
to them also that at the first requisition of the patentee of these
presents they shall do [illegible] respective, and to execute
during their duration as if it were the Law of the Kingdom.
In witness of which we have signed and had the said presents
counter-signed, to which we have had affixed the seal of the
State, at Paris the twenty ninth day of the month of November, one thousand seven hundred and ninety one and the
eighteenth year of our Reign.

(Signed) LOUIS

by the King

The Minister of the Interior

B. C. Cahier.

LEGEND

*A*, the large piston acting upon the wheel *B* which gears with the wheel *C*, and makes the wheel *D* turn continuously towards the supports where the large piston always moves from above downwards, and acting on the endless chain *aaa* makes the wheel *E*, fixed on the extremity of the axle, turn continuously towards the supports.

The arms *CCC*, placed on the quarters of the boat, support the ends of the axle, and the framing *CCC* supports the ends of the paddles by means of arms suspended by the hinges *ddd*. The paddles are so constructed that the end of the axis turns into the wheel, and in proportion as it turns it makes the paddles *ggg* move somewhat as a man puts them in motion in a boat.

N.B. This drawing represents the view of the works constructed on the stern of the boat, the perspective of the large piston, and one of the sides of the boat.

At the foot of the drawing is the following: "Brian architect and draughtsman of the Directory of Patents of Invention."

In 1793 Fitch himself crossed to France, but the French Revolution was then reaching its climax and he soon returned home a disappointed man, working his passage as a seaman. Back again in America he experimented with a screw-driven steam boat on the Collect Pond, New York, but in spite of all his ingenuity and perseverance he failed to make any headway, and overcome by his misfortunes he became ill and in the summer of 1798 died at Bardstown, Kentucky. His position as an original and independent inventor is unassailable and whatever admiration is accorded him is increased by a realisation of the backward state of engineering in the United States in the eighteenth century. He was a very far-sighted pioneer and once wrote that with steam "The Grand and Principle object must be on the Atlantick, which would soon overspread the wild forests of America with people, and make us the most opulent Empire on Earth". Though Fitch's services to mankind were long overlooked, in recent times the States of Connecticut and Pennsylvania, the town of Trenton, and finally the United States Congress have each in some way commemorated his achievements.

The experiments of Miller and Symington in Scotland were made almost simultaneously with those of Rumsey and Fitch, but whereas Rumsey and Fitch were rivals and intensely jealous of each other, Miller and Symington were collaborators and stood in the relation of employer and employee.

Patrick Miller (1731–1815) was an Edinburgh banker with novel ideas on ships. He entertained the view that ships with twin hulls would prove superior to those with single hulls; he made several such double-hulled boats and also one, the *Edinburgh*, with triple hulls. These were all sailing vessels, but to some of them he fitted paddle-wheels driven by man-worked capstans. The suggestion that steam should be used instead of man-power was apparently made to Miller by James Taylor, the family tutor, and this led to the collaboration of William Symington (1763–1831). In 1787 Symington, who was the son of the engineer of the Wanlockhead Mines in Lanarkshire, took out a patent for a "New Invented Steam Engine on principles entirely New". Some of the ideas embodied in the engine were not so entirely novel as one might be led to expect, for in his method of condensing the steam Symington infringed the famous patent of Watt, who later on obtained an injunction against him. Symington however built an engine for Miller and on October 14, 1788, it drove one of Miller's boats on Dalswinton Loch, Dumfriesshire, at the rate of about four miles an hour. Next year a second and larger engine was made and tried, but Miller was far from pleased with the results and wrote, "I am now satisfied that Mr Symington's steam engine is the most improper of all steam engines for giving motion to a vessel." Of the second engine nothing remains, but the engine of 1788, which for long stood in Miller's library, is preserved in the Science Museum, South Kensington. With its two open-topped cylinders, 4-in. diameter, with air pumps and condensers below, and a strange arrangement of rods and chains, pulleys and ratchets for obtaining rotary motion, it is an object of great historical interest. It is not now in its original state for, according to a letter in *The Engineer* for July 14, 1876, the engine before being placed on exhibition underwent extensive restoration by John Penn.

Symington's prominence among the pioneers of the steam boat, however, is not so much due to his work for Miller as for the construction of the engine for the *Charlotte Dundas*, built at Grangemouth in 1801. On October 14, 1801, Symington patented his "New Mode of Constructing Steam Engines and applying their power of providing rotatory and other motions without the interposition of a lever or beam", and his engine for the *Charlotte Dundas* was designed in accordance with this patent. The engine had one double-acting horizontal cylinder 22-in. diameter, 4 ft. stroke, the piston rod of which drove the crankshaft of the stern paddle-wheel directly through a connecting rod. The simplicity and superiority of this engine was evident. Many successful trials were made with the *Charlotte Dundas*, and Symington's prospects appeared very bright, but it was all to prove in vain. Circumstances occurred which led to the *Charlotte Dundas* being laid up and

Symington from that time almost passed out of history. When over sixty years of age he appealed to the Government for financial assistance and was awarded two meagre sums of £100 and £50. Sinking under his misfortunes he became dependent on relations and died at his son-in-law's house, No. 44 Burr Street, near St Katherine's Dock, London, and was buried in the churchyard of St Botolph's Church, Aldgate, not far from where now stands the premises of the Institute of Marine Engineers. Years afterwards the simple type of direct-acting engine he made for the *Charlotte Dundas* became the standard form for marine engines, and no one deserves the title of the "Father" of marine engineering more than Symington, who did for the steam boat what Trevithick did for the locomotive.

The next definite advances were made in America by Colonel John Stevens (1749–1838) and by Robert Fulton (1765–1815). In 1804 Stevens fitted the *Little Juliana*, 24 ft. long, with a high-pressure multitubular boiler supplying steam to a single-cylinder non-condensing engine driving twin screws. The boat was navigated in the harbour of New York in 1804, at an average velocity of four miles an hour. Stevens had laboured for several years at the introduction of screw propulsion, but like all American inventors of his time was much handicapped by the lack of tools and competent workmen. To him we owe the first steam vessel to make a passage in the open sea, the paddle wheel *Phoenix* of 1808. Robert Livingston Stevens (1787–1856) and Edwin Augustus Stevens (1795–1868), sons of Colonel Stevens, were both pioneers of steam navigation in the United States.

Contemporary with the work of Colonel Stevens came that of Robert Fulton—"An Artist of high merit, a Civil Engineer of ability, a social philosopher of deep insight and warm affection." When in France in 1803 Fulton experimented with steam boats on the Seine; he afterwards saw the *Charlotte Dundas*, and before returning to America in 1806, after an absence of nineteen years, he ordered an engine from Boulton, Watt and Co., of Birmingham, and a boiler from London, and these he placed in the historic *Clermont*, popularly called "Fulton's Folly". It was with the trip of the *Clermont* on August 17, 1807, up the Hudson River, that steam navigation as a regular means of transport began. In H. W. Dickinson's *Robert Fulton, Engineer and Artist* will be found a full account of the *Clermont* and of the other steam boats Fulton constructed, including the *Demologos*, the first of all steam fighting ships. Fulton's success stimulated others to follow his example and soon steam boats were to be found on all the great waterways of the North American continent.

Just as it was Fulton's *Clermont* which inaugurated steam navigation in the New World, so it was Bell's *Comet* which inaugurated it in the Old World. Henry Bell was born at Torphichen, near Linlithgow, April 7, 1767, and died at Helensburgh, November 14, 1830. A carpenter by trade, he had as early as 1801, and again in 1803, suggested to the Government the application of steam to boats. In 1808 he became the proprietor of a hotel and baths at Helensburgh, and it was with the object of conveying his guests to and from Glasgow, about fourteen miles distant, that he had the *Comet* built. Charles Wood built the boat, John Robertson made the engine and David Napier the boiler. She was launched in July 1812, and a century later the *Glasgow Herald*, in which her sailings had first been advertised, published an interesting centenary number containing an account of her career. Bell, like Fulton, found many ready to make use of his experience, and the story of the *Comet* and her successors is well told in Captain James Williamson's *The Clyde Passenger Service*. Superior vessels soon drove the *Comet* from the Clyde but Bell gallantly struggled on, placing her now on this service and now on that, until on December 15, 1820, the famous little craft was driven on the rocks at Crinan. By chance the forward part of the boat held together, the engine was salved, and after doing duty in a Glasgow factory it was presented to the nation through the generosity of Robert Napier and Sons, and now stands in the Science Museum beside Symington's engine of 1788.

Like most of the other pioneers Bell reaped no adequate financial reward and he was never able to pay Napier the full price of the boiler. In the closing years of his life friends came to his assistance and he was given a pension by the Trustees of the Clyde Navigation. His grave, surmounted by his statue, is in Row Churchyard near Helensburgh; at Helensburgh itself is a memorial obelisk to him, while yet another obelisk stands on the banks of the Clyde, at Bowling, recalling to those who go down to the sea in ships his great gift to human progress.

## CHAPTER II

# EARLY PROGRESS OF STEAM NAVIGATION

THE work of Fulton and of Bell definitely marked the inauguration of the use of steam boats on the rivers of America and Europe respectively. This was an innovation for which the world had been prepared in the previous century by the increased use of machinery, and there were many engineers ready to try their hand at making marine engines and boilers. From the Hudson and the Clyde the new means of transport spread to the St Lawrence, the Mississippi, the Great Lakes, the Mersey, the Thames, the Seine and the Rhine, and even to India and to Batavia. Perhaps the most extensive collection of facts concerning early steam boat enterprises is contained in Preble's *Chronological History of the Origin and Development of Steam Navigation*, published in Philadelphia in 1883. For 25 years Rear-Admiral George Henry Preble, U.S.N. (1816–85), collected notes on the subject, and his book is a veritable storehouse of information. From such figures as are given by him and others, it appears that by 1819 about 100 steam vessels had been constructed in the United States while, in 1820, the British Empire possessed 43 vessels of the total tonnage of 7240 tons. Twenty years later, the tonnage belonging to the United Kingdom was 95,807 tons, while that of the United States was 200,000 tons. Of the latter, only about 4000 tons was registered for foreign trading. While Great Britain was building up a fleet of steam vessels suitable for trading in any waters,

American constructors were mainly engaged with river craft, and it has been said that almost every venture made with American steamers upon the ocean during the thirty years succeeding the *Clermont's* first trip on the Hudson proved unprofitable.

If American shipbuilders and marine engineers found full scope for their skill in meeting the demands of inland transport, it was otherwise in this country. Our geographical position, no less than the needs of our far-flung Empire and our great overseas trade, naturally led to the building of vessels fit for work in the open sea, and it is with the development of such ships that this history is mainly concerned. In this chapter is given a résumé of some of the work of the two decades preceding the establishment of transatlantic steam navigation.

Before proceeding further, it should perhaps be pointed out that, though steam navigation became of increasing importance, for many years steam tonnage was small in comparison with sailing tonnage, and long after the advent of the steam ship our sailing tonnage continued to increase. The relative and approximate figures in 1840 were: steam vessels, 95,000 tons, sailing vessels, 3,000,000 tons; in 1849, steam vessels, 170,000 tons, sailing vessels, 3,000,000 tons; in 1866, steam vessels 747,000 tons, sailing vessels, 4,705,000 tons; in 1871, steam vessels, 1,290,000 tons, and sailing vessels, 4,343,000 tons. Until the adoption of the compound steam engine, the sailing ship remained the most economical means for carrying large cargoes long distances and, under favourable circumstances, the splendid clipper ships could outdistance the finest steam ships. In the 'eighties, however, our steam tonnage was equal to our sailing tonnage, and each succeeding decade has been marked by a rapid decline of sailing tonnage, until, to-day, the construction of sailing vessels of any considerable size has ceased. It may be that the marine engineer has destroyed much of the romance of the sea, but he has certainly diminished the dangers and delays of sea transport.

Among the best known marine engineers in this country in the first half of the nineteenth century were Robert Napier, David Napier, Scott, Sinclair, Caird and Tod of the Clyde; Fawcett, Bury and Forrester of Liverpool; and the Maudslays, Field, Miller, Barnes, Seaward, Rennie and Penn of the Thames. The historic firms of Boulton, Watt and Co., and the Butterley Iron Works, Derbyshire, also built marine engines, and these firms, together with that of Maudslay, Sons and Field, were at one time considered the only ones of sufficient reputation to execute Government contracts. The early experimental steam boats had engines of various designs, but not many years passed before the Watt beam engine was modified by having beams placed on either side of the cylinder and crank, instead of overhead; known as the side-lever engine, this type was familiar to every sea-going engineer. It was in its day the standard type, and as well-known as was the vertical triple-expansion engine of more recent times. Every maker constructed such engines and the first cross-Channel vessels, the first naval steam vessels, the first steam fighting ships, and the earliest ships of the General Steam Navigation Company, the Peninsular and Oriental Steam Navigation Company, and the Cunard Company all had side-lever engines. On a stone plinth in the grounds of Dumbarton Castle is to be seen the single-cylinder side-lever engine made in 1823 by Robert Napier for the *Leven*. This engine was of only

33 nominal horse-power, but nearly forty years later Napier was constructing engines of a similar type for the last paddle-wheel vessel of the Cunard Company, the *Scotia*, the engines having two cylinders 100 in. diameter and 12-ft. stroke. Supplied with steam at 20-lb. pressure, these engines developed 4900 indicated horse-power. American engineers, for many of their river boats, used engines with overhead beams, but they also constructed side-lever engines, some fine specimens being found in the ships of the Collins Line, which attempted to wrest the supremacy of the Atlantic from the Cunard Company.

At first, steam pressures were very low—little above atmospheric pressure—and the boilers were simply great rectangular boxes with square furnaces and long winding flues of such a size that a man could pass through them. They were sometimes made of iron, sometimes of copper, and, of course, were nearly always worked with salt-water.

A steam vessel which may be taken as representing the best practice of the day was H.M.S. *Lightning*, built at Deptford in 1823 and engined by Maudslays. She was 126 ft. long, 22⅔ ft. wide and 296 tons. Her engines, of 100 nominal horse-power, had two cylinders 40½ in. diameter and 4-ft. stroke. During her career, which lasted till 1872, she had four sets of boilers but retained the original engines, and of these there is a sectional model by Henry Maudslay in the National Maritime Museum at Greenwich. Her original flue boilers carried but 2 to 3-lb. pressure. To the *Lightning* belongs the distinction of being the first steam vessel in the British Navy to take part in naval warfare, and the first for the command of which a commission was granted. In 1824, she accompanied the bomb vessels to Algiers; the following year she was being fitted out as a yacht. On December 4, 1827, the Duke of Clarence signed the commission of Lieutenant G. Evans for her command, and a few years later she is referred to in *Punch* when recording an incident connected with the crossing to Ostend of a royal personage. *Punch* then printed the lines:

> Though Canute could not check the wave
> When rapidly the tide was heightening,
> Victoria her orders gave
> And at her bidding stopped the *Lightning*.

In a picture of an incident in the Baltic campaign of 1854 the *Lightning* is seen leading the *Edinburgh, Hogue, Amphion, Alban, Blenheim* and *Ajax* to the attack on Bomarsund, and towards the end of her career she was lent to the Royal Society for deep-sea dredging. Interest in the vessel is not lessened by the existence of a letter from her engineer, written in 1826, to Simon Goodrich, the Engineer of Portsmouth Dockyard. This letter, which recalls the constant anxiety of the early sea-going engineers regarding the density of the water and their method of testing by boiling, is probably the earliest letter extant from any engineer of a steam vessel in the Navy. From what has been said of Maudslay (whose name in the letter is spelt Mosely), and of the trip to Algiers, the references will be understood. The letter is from John Chapender; it is given here in full.

Deptford, August 1, 1826.

H.M. Steam Vessel Lightning.

Sir,—I should have wrote to you before of our arrival at Deptford but have been so bussey imployed in towing ships about that I have had not a hour since I left Cronstead in Rusia we left in on the 1 of July in com-

pany with the Glouester 74 after beatin and towing her
at time we was ten days and nights before we made
Copheagen after working that time my water in the
boilers did not exceed the tempeture 216 Deg. by my
blowing some water from the boilers several time a day.
We took our departure from Copheagen on the 13th
inst. and then whent into a port in Norway in a gale of
wind on the 17 Inst at this place Biggs Departed on 19
Inst this life and was buried their the tempeture of the
water in the boilers did not exceed 216 Deg. took our
departure from Egersound in Norway on the 21 Inst
and made Sheerness on the 24 Inst which made our
runing better then two hundred miles in four and
twentey hours when We came to Sheerness we run to
Chatham and then towed a 74 from Sheerness to near
Chatham left Sheerness for Deptford on the 24 Inst and
then the next day towed a ship from Woolwich to the
Nore and then back to Deptford to morrow we towes a
barge from this to the Downs or near Portsmouth with
the Embasadore things that his goine out in the Ganges
and then return back to Deptford immedently Sir the
tempeture of the warter in the North sea his 214 and
the tempeture in my boilers did not excead 218 so that
I never put my fire out nearly the old of the voage and
when I took my man hole of the boilers of was as clean
as a pair of new ones my Engins his in as good repair as
when I left England the main pipe joint under the
boilers give away but I repaird that so that it his as tite
as ever but my front plates round the fire doors his
gitin bad and my Cluch on the paddle Wheels his git
lose by been in such evey seas. Mr. Mosely have been
on board and it give him grate pleasure to see the
Engins in so good a state after such a long journey Sir
I hope that you will send to the Board to know if I am
to have the boat or know as we belong to your department for I have had know time to see them or
write to them I hope that you will do me that favour
and let me know Sir as soon as posable or some Recom-
pence for bring home the Vessel as I am at Deptford I
have two or three places to go to if the don give me the

situation. I have had charge of the Engins from the 4 of June when we left Cronstead to look for the Duke I have had a fertigen time of it Algers was nothing to compair to this Journey.

Sir,

I remeins your

most humble and

Obedgent Servent

JOHN CHAPENDER, Acting Engineer

of H.M. Ship Lightning.

The original of this letter is preserved in the Science Museum Library with the Goodrich papers. It is accompanied by another letter dated November 8, 1826, in which Chapender acknowledges with gratitude his appointment. But Chapender evidently did not give satisfaction for, in Goodrich's note-books, are found the entries:

5. 5. 27—Sat. at Dyd [Dockyard]. Writing about Chapender. .

7. 5. 27—Send on report for the removal of Chapender from the Lightning.

11. 5. 27—Further explanation about Chapender.

7. 6. 27—Recommend Jenman for 1st Engineer of the Lightning.

Early marine engineering practice is also illustrated by the story of the *Enterprize*, the vessel by which it was hoped to establish steam communication with India. Meetings of officials and merchants had been held in London in 1822 and in Calcutta in 1823, to consider the project, and the Government of India had offered 20,000 rupees to any British subject who would permanently establish steam communication either via the

Cape or the Red Sea before the end of 1826. Funds being forthcoming, the *Enterprize*, then on the stocks of Gordon and Company, Deptford, was purchased and Lieutenant J. H. Johnston, R.N. (1787–1851), one of the chief promoters, was chosen to command her. The *Enterprize* was 122 ft. long on the keel, 27 ft. beam, and 479 tons burden, and into her Maudslays fitted one large copper flue boiler and a two-cylinder side-lever engine, with cylinders 43 in. diameter and 4-ft. stroke, driving paddle wheels 20 ft. in diameter. Gear was supplied for holding the wheels when the ship was under sail. In the engine room were brine pumps for discharging water from the boilers and "refrigerators" for heating the incoming feed water by the outgoing brine. Much of the coal was carried in iron tanks which, when emptied, could be filled with sea water to ensure a proper immersion of the paddle wheels. The vessel cost about £43,000.

Though often referred to, no full account of the voyage has been published, but in the Field papers, presented a few years ago to the Science Museum Library by Miss Gertrude Field, a granddaughter of Joshua Field, are a copy of Lieutenant Johnston's log up to the time that the vessel arrived at the Cape, and a letter written from Calcutta to Henry Maudslay by William Ash, the engineer of the ship.

Leaving Falmouth on August 16, 1825, with seventeen passengers and a small quantity of cargo, the ship reached Cape Town on October 13, and Calcutta on December 7. Though she did not fulfil the conditions necessary to obtain the reward, the *Enterprize* was purchased by the Indian Government, who employed her in running despatches between Calcutta and Rangoon, and she is heard of again when the mail service to India via Alexandria and Suez had been started. The log is too long to be reproduced, but one or two extracts will help to illustrate the problems that confronted the commanding officer. During his two months' voyage to the Cape, Lieutenant Johnston had many anxieties and difficulties to contend with, some of which arose from the distribution of coal in tanks in various parts of the ship. For days together the ship must have been like a collier, but both passengers and crew bore "with these disagreeables with much consideration". The first entry reads:

Tuesday, August 16, 1825. Draft of water forward 15 ft. aft 15 ft. 3 in. At 7.30 cast off from Mooring Buoy and proceeded down Channel. 9.30 passed the Lizard Light. Engines going 22 strokes. Temperature 224 deg. Speed 6½ miles per hour. Light breezes from the north-west.

On August 27 the ship, with the aid of steam and sail, had got as far as the Canaries, and on that day the fires were worked down, the engines stopped, seven paddle boards aside were taken off, the wheels were locked, and the vessel proceeded under sail. Johnston's log then records:

August 28. The engines have been continually at work since the 16th inst. at noon, a period of 11 days, during which time the water has only been blown off once, and then more as a measure of precaution than of necessity. The engineers and the stokers have been much fatigued and, to-day being Sunday, as we are, I trust, fairly in the Trade, I have desired everything connected with the engineer's department to stand over till to-morrow. I have invited the chief engineer, Mr. Ash, to dine with me, as I wish to give him as much consequence as possible in the eyes of those who are placed under him.

August 29, a.m. Opened the boiler found all the water had not blown off, and incrustation on the bridges and on the chimney casing, of and about the thickness of a half-sovereign is the only deposit in the boiler. This is confined to the bridges and to the sides of the fireplaces as low down as the bars; below the bars the copper is bright. The stokers are employed cleaning it off with sharp-edged tools, which cut the copper and which, unless much care is taken, must injure it considerably. I find that a moderate blow of a smooth-faced hammer will remove this incrustation more quickly, the concussion having the effect of detaching it in large flakes. Ash objects to this mode from an idea that the repeated blows will expand the copper and start the rivets. This I do not think will be the case, as the blow required is not I think more than a force of 2 pound on a square inch of surface. The water remaining in the bottom of the boiler contains a considerable quantity of magnesia, and is more turbid than what was delivered by the refrigerator at the last moment. Engineers about the engines lifting the covers of the cylinders and adjusting the pistons. Sweep cleaning the flues found the flue had wept very much. Saved the cow dung to plaster the bridges, &c. I have examined the coal tanks, and I find the consumption has been considerable....I think certainly that more than one-third of the whole quantity of coals has been expended, and I shall be tenacious of using the steam when I can go 5 or 6 knots with sail.

After being cleaned, the boilers were again lighted, and with the expenditure of an extraordinary amount of labour in transporting coal, steam was maintained and the engines used off and on till the ship reached the island of St Thomas on September 17. From that time till the arrival at the Cape the entries are much shorter, and refer mainly to the navigation of the ship. The last two entries read:

October 12, 7 p.m. Lit the fires, furled the sails, and

set on steam. At midnight hove to off Table Mountain bearing S.E.

October 13. At daylight set on steam. Hoisted the anchors from the fore hatchway, and got them over the sides. Got the cables up and stood into Table Bay. Received a salute from the Castle and anchored in 5 fathoms.

The letter of Ash to Henry Maudslay bears no date, but it was evidently written some time after the vessel had been taken over by the Indian Government. He referred to the breaking of the brass bolts in the boiler, to trouble with the discharge pipe and cock, to broken spindles, to valves out of order, to repairs of the "fregerator", "but after all these trifles we have done very well and to satisfaction, as now it is said from good authority that the vessel has showed herself by her great services". Ash also had something to say of the little *Diana*, which had been used in the Burmese campaign of 1824, adding "the engineer, Mr. Darwood, that came with her has lost his leg by getting under one of the beams in a seaway, but is about as well as he can with an assistant but talks of coming to England"; further on he says, "The *Falcon* Steam Vessel that the *Enterprize* was not to be compared with in England has arrived here after about 4 months passage now they have her engines out and made a vessel for cargo of her. The Engines are now lying on shore."

At the time the *Enterprize* made her voyage, there were many shipping companies owning steam vessels and others which were concerned only with such vessels. Of the latter, two of the best known were the City of Dublin Steam Packet Company of Liverpool, and the General Steam Navigation Company of London, which, in 1826, sent a vessel as far as Lisbon. The

Admiralty also acquired other steamers beside the *Lightning* and *Comet*, and by 1830 the names of about a dozen appeared in the Navy List. To some of these fell the task of first carrying mails overseas, and the service inaugurated by the Navy was the first regular series of long-distance voyages made under steam.

Mails had been sent to France by steam boat since 1821, but, up till 1830, letters for more distant countries were carried by sailing packets, the terminal port for which was Falmouth. For more than a century, and until after the Napoleonic wars, these sailing packets were managed by the Post Office. A year or two after peace was declared, however, the service was taken over by the Admiralty, the vessels forming an admirable training ground for seamen. Falmouth, in those days, held a position much the same as Southampton holds to-day. It was in connection with the conveyance of mails that the oft-quoted minute of Lord Melville, the First Lord of the Admiralty, was written. In 1828, Mr Hay, of the Colonial Department, had asked that a steamer might carry the mails from Malta to Corfu. To this, Lord Melville, replying in a minute written by himself,

regretted the inability of my Lords Commissioners to comply with the request of the Colonial Department, as they felt it their bounden duty, upon national and professional grounds, to discourage, to the utmost of their ability, the employment of steam vessels, as they considered that the introduction of steam was calculated to strike a fatal blow to the naval supremacy of the Empire, and to concede to the request preferred would be simply to let in the thin end of the wedge, and would unquestionably lead to similar demands being made upon the Admiralty from other departments.

It is probable that the official view reflected the general opinion of seafaring folk, for at that time we owned but 293 steam vessels as against some 20,000 sailing vessels. Steam boats, however, had proved themselves so much faster on short journeys and maintained so much greater regularity that, in 1830, it was decided to send letters to the Mediterranean by steam, and it thus fell to the lot of the Navy to inaugurate the long-distance overseas steam mail service. The first vessel to be put on the run was H.M.S. *Meteor*, of 296 tons, and 100 nominal horse-power, and her departure from Falmouth on February 5, 1830, marks an epoch in the history of steam navigation. Sailing ships had taken three months to do the round voyage to Corfu and back, but this time was speedily reduced by the *Meteor* and her consorts, and, in 1831, the *Messenger* performed the outward and homeward passages in thirty-one days. The vessels first employed were the *Echo*, *African*, *Carron*, *Confiance* and *Columbia*, which had all been built in the Royal Dockyards, and to these were added the larger vessels *Hermes*, *Messenger* and *Firebrand*, the two former of which had been built at Blackwall for the General Steam Navigation Company, as the *George IV* and *Duke of York*. These two were considerably larger than any of the others, being of 733 tons. The engines of the vessels were constructed either by Boulton, Watt and Co., the Butterley Iron Works, or Maudslays. The boilers were the rectangular flue type, of iron or copper, and all the vessels had side-lever engines. Though few seafaring engineers of those times left any account of their experiences, thanks to John Dinnen (1808–66), the first naval engineer to obtain the rank of Inspector of Machinery Afloat, who for five years was chief engineer of the *African*, we have an ex-

cellent review of the practice of the time, while one of
the commanding officers of the day, Commander Robert
Otway, published an *Elementary Treatise on Steam—as
Applicable to the Purposes of Navigation*; the first edition
of Otway's book appeared in 1834; Dinnen published
his memoir on marine boilers in 1838, and both are
invaluable to the student of marine engineering history.
Though Otway said he found the engineers he met
"almost invariably solicitous to evade, by technicality
of language, imparting information on the subject; and
in fine, to use every endeavour to make a mystery of the
whole operations of the engine room", his book is full
of practical details relating to the construction and
manipulation of the engines. Otway was a believer in
higher pressures; he recommended that the Royal
Dockyards should undertake the repair of engines and
boilers; and argued that if the Government wanted the
right men they would have to start a training school
for engineers. He told his brother officers that, though
they should be acquainted with the principles of the
machinery, it was no place of theirs to superintend the
engine room or to crawl through a boiler, "all that is
required is commonplace prudence and attention, and
neither too implicit confidence in those immediately
attendant on the engines, nor too rash an application
of steam for the mere purpose of showing off, by an
acceleration of speed, or the very censurable competition with a rival steamer". The boilers he regarded as
"the fountain head from which all benefit is derived",
but he complained that vessels running mails were not
allowed proper time to clean the boilers, and the commanding officers sometimes fell under censure. In another passage, when speaking of the work to be done in
the engine room on arriving in harbour, he gave some

good advice to his brother officers as regards their attitude towards the chief engineer, remarking "furthermore, by showing this deference and mark of respect to that officer it serves as an example to the men on board who will the more readily look up to him as their superior in every point of view".

Dinnen's memoir was naturally more concerned with the purely technical side of the subject, and it may be recalled that, at a time when it was supposed that the Mediterranean was much salter than the ocean, Dinnen, by blowing down a portion of the water in the boilers of the *African* every two hours, kept them practically clean and free from marine deposit, "a circumstance without precedence". The boilers of the *African* were of copper and the leakage of salt water into the flues gave Dinnen far more trouble than incrustation. Even then he saw that "nothing but the use of distilled water will ever thoroughly obviate the evils to which marine boilers are subject on very long voyages". He said nothing of those above him, but in speaking of economy of fuel, he remarked: "Stokers consult their own convenience; and as long as they keep plenty of steam flying off to waste they are seldom called to account; as to myself I have found the greatest difficulty in obliging them to keep their fires uniformly."

The Navy maintained the mail service, by means of the vessels mentioned, until 1837, when a contract was made with the Peninsular Steam Navigation Company, and, in September the same year, the *Iberia*, of 400 tons, built by Curling and Young, and engined by Miller and Barnes, sailed for Gibraltar, calling at several ports in Spain.

With the extension of the contract to other ports in the Mediterranean, the activities of the Company increased, and under the title, "Peninsular and Oriental Steam Navigation Company", or more familiarly the "P. and O. Company", it has ever since carried on and extended the work begun by the Royal Navy. If it were possible to continue the story of the extension of the mail service to the East, reference would have to be made to the use of the overland route, the navigation of the Red Sea, first by the *Hugh Lindsay*, then by the *Enterprize* and the *Atlanta*, till we come to the *Berenice*, which Robert Napier supplied to the Indian Government. This vessel may be regarded as Napier's finest piece of work prior to his construction of the first vessels of the Cunard fleet. Built in 1836, the *Berenice* was 170 ft. long by 28·8 ft. broad and of 646 tons, with engines of 250 nominal horse-power. An account of her voyage to India is given in the *Life of Robert Napier*. She left Falmouth on March 16, 1837, and reached Bombay on June 13. She was under steam sixty-three days, but was detained no less than twenty-five days coaling at various ports. Her average speed for the whole voyage was 8 knots, and it was largely her success which led to the co-operation, two years later, of Napier with Samuel Cunard. Some particulars of the *Berenice's* voyage are given below:

| Passages | Distance miles | Coal expended tons | Highest speed knots |
|---|---|---|---|
| Falmouth to Santa Cruz | 1510 | 128 | 10 |
| Santa Cruz to Mayo | 896 | 87 | 11 |
| Mayo to Fernando Po | 2284 | 190 | $9\frac{3}{4}$ |
| Fernando Po to Table Bay | 2269 | 220 | 10 |
| Table Bay to Port Louis | 2488 | 313 | 10 |
| Port Louis to Bombay | 2612 | 170 | 10 |

The *Berenice*, unlike the *Enterprize*, found coaling

stations available. Notwithstanding this, her voyage was a notable one, and it marked the limit of achievement by steam until a year later, when the grand project of transatlantic steam navigation was shown to be practicable by the famous *Great Western*.

PLATE II

*GREAT WESTERN* PASSING PORTISHEAD, APRIL 8, 1838
From contemporary lithograph. Courtesy of Science Museum

MODEL OF *BRITANNIA*
Courtesy of Science Museum

## CHAPTER III

## PIONEER TRANSATLANTIC STEAMSHIPS

ONE of the great landmarks in the history of the steamship is the inauguration of transatlantic steam navigation. It was the Atlantic traffic which called into being the large and powerful steam vessel, and in that traffic have always been found the finest and fastest vessels afloat. For a century now there has been regular steam communication between the Old World and the New World, and what was at first a purely British undertaking has become a great international service. America, France, Germany, Holland, Belgium, Italy and other countries are all engaged in this important enterprise, which was started with British capital and British ships alone. The early vessels of not much more than 200 ft. in length, with engines of a few hundred horse-power, have given place to ships five times as long, and fitted with machinery equal in size and power to that found in our super-power stations ashore. Though in the present century there have been many remarkable improvements in the ships of the Atlantic service, the birth and development of that service will always remain as one of the greatest achievements of the Victorian age.

If to Great Britain belongs the credit of inaugurating transatlantic steam navigation, it is to the American pioneer, John Fitch, that we owe the first conception of the possibilities of such traffic. Fitch died in 1798, and forty years elapsed before the *Sirius* and *Great Western* made their first passages to New York, yet he appears to have been alone in visualising such passages. To

America also belongs the distinction of first despatching a steam vessel across the Atlantic. In 1819, twelve years after Fulton had launched the *Clermont*, the little *Savannah*, of something over 300 tons, made a voyage from America to Europe. Launched as a sailing vessel, she was afterwards fitted with a single-cylinder steam engine and with paddle wheels which could easily be unshipped. A model of her is in the National Museum, Washington, and the story of her voyage is given in Spears' *The Story of the American Merchant Marine*, and in other works. Leaving Savannah towards the end of May 1819, she arrived at Liverpool on June 20, having during a passage of some 27 or 29 days used her engine for 80 hours. From England she went to Elsinore, Stockholm and St Petersburg, and then returned to America where her machinery was removed. She then traded as a sailing vessel until wrecked in 1822. Her size, her limited power, and the small amount of fuel she carried made it quite impossible for her to steam any considerable distance, and her voyage to England exceeded in length those of many of the sailing packets. She is constantly referred to as the first steam vessel to cross the Atlantic, but that statement is apt to give a wrong impression. Whatever interest there is in her voyage, the decision to send her to Europe was not connected with any considered scheme for regular steam passages, and it had no influence on the early projects for transatlantic steam navigation.

Nor was the *Savannah* the only steam vessel to cross the Atlantic before 1838. There was never any great difficulty in sending small steamers to distant stations under sail, or with the engine used as an auxiliary source of power. The *Curaçoa*, for instance, a vessel of 436 tons, built at Dover in 1826, after being purchased by the Dutch Government made three voyages during 1827–29 between Holland and Dutch Guiana, taking about a month for each passage. To have steamed the whole way she would have needed more than 200 tons of coal. Of another voyage made partly under steam, partly under sail, we have more detailed particulars in the log of H.M.S. *Rhadamanthus*, preserved in the Public Record Office, London. The first steam vessel built by the Admiralty at Devonport, the *Rhadamanthus* was of 800 tons and 200 nominal horse-power. Launched on May 16, 1832, she was employed for a time with her sister ship the *Dee*, in the blockade of the Dutch coast, and was then sent to the West Indies. Leaving Plymouth, April 21, 1833, she steamed across the Bay of Biscay and then, having stopped her engines and unshipped six paddle floats from each wheel, she continued her voyage to Funchal under sail. A coal brig was already awaiting her at Madeira, and after replenishing her bunkers, the *Rhadamanthus*, on April 30, left on her voyage of 2500 miles to Barbadoes, which was reached on May 17. Steam was used intermittently, the floats of the paddle wheels being removed and replaced as necessary. The log contains a few references to the working of the engines, and on more than one occasion notes the issue of an extra allowance of spirits to engineers, stokers and coal trimmers.

The voyage of the *Rhadamanthus* was from east to west, and in a region of fine weather, but a few months later, the *Royal William* crossed from west to east in much stormier latitudes. Built at Quebec, 176 ft. in length over all, and of something over 800 tons, this vessel had an engine of 200 nominal horse-power, constructed by Bennet and Henderson of

Montreal. Launched in April 1831, she traded for a time between Quebec and Halifax, but business being bad, her owners decided to send her to England to be sold. She accordingly left Quebec on August 4, 1833, called at Pictou for coals, crossed to Cowes, Isle of Wight, and arrived at Gravesend, September 11. It is sometimes stated that she steamed all the way, but this is difficult to reconcile with the time she was on passage. In her case, the fuel problem was solved by her carrying coal as cargo, but the details of her voyage do not appear to be known. After arrival in England she was sent to Bordeaux, and as the *Ysabel Secunda* sailed under the Spanish flag. The particulars and a model of the *Royal William* are preserved by the Literary and Historical Society of Quebec.

Though the voyages of the *Savannah, Curaçoa, Rhadamanthus* and *Royal William* did not demonstrate the practicability of the navigation of the Atlantic by ships using steam as the main motive power, as early as 1825 a company had been formed for running steam vessels between Ireland and Nova Scotia, and between Ireland and the West Indies. The project was commented on favourably by some of the newspapers of the day, but, as the vessels it was proposed to use were only of 500 tons, there would have been little chance of success had the company begun operations. At that time the project was certainly a bold one, but it was supported by sound arguments and one of the pamphlets advocating it said:

It may be assumed as an uncontrovertible fact that wherever steam navigation has been established on a proper footing and on a sufficient scale of vessels and machinery, it has not only been abundantly successful, but its performance has surpassed expectations, overcome the natural prejudices and commanded the confidence of even nautical men, and it has not only drawn to it all the most valuable communication in its line of transit, but also increased it in a tenfold proportion.

Some years, however, had to pass before the vessels and machinery of adequate scale were available, and then, in the memorable year 1838, three British steamship companies almost simultaneously despatched steam vessels to New York. With those three companies and their ships the proper history of the "Atlantic Ferry" begins. If it is desired to credit transatlantic steam navigation to any one individual, it is to the American lawyer and man of business, Junius Smith, that such credit is due. Smith, who was born at Plymouth, Massachusetts, in 1780, had a successful career at the American bar and then engaged in business in London. In August 1832 he paid a visit to the United States, crossing in the *St Leonard* and returning in the *Westminster*, two of the sailing packets. The voyage to New York took fifty-four days, the return voyage to Plymouth thirty-two days, a fair example of the tedious nature of Atlantic voyages about a century ago. In the *Westminster* Smith began to ponder over the idea of placing steamers on the route, and six months after his arrival in England he wrote: "Thirty-two days from New York to Plymouth is no trifle; *any ordinary sea-going steamer would have run it with the weather we had in fifteen days with ease.* I shall not relinquish the project unless I find it absolutely impracticable." Fortunately, with his foresight, Smith possessed determination, and, in spite of opposition and ridicule, in 1835 he was able to float a company and in 1836 he again wrote to New York, "I have the pleasure to inform you that the Directors of the 'British and American Steam Navigation Company' have contracted for the building of the largest and intended to be the most splendid steamship ever built, expressly for the New York and London trade. She will measure one thousand seven hundred tons, two hundred feet keel, forty feet beam, three decks, and everything in proportion. She will carry two engines of two hundred and twenty-five horse-power each, seventy-six inch cylinder, and nine feet stroke. The expense of this steam frigate is estimated at 60,000£. These large undertakings will require time to mature, but I think the business will at last be done effectually". The vessel Smith referred to was constructed by Curling and Young, Limehouse, who had just completed the *Iberia* for the Peninsular Steam Navigation Company, and had the engine makers, Girdwood and Company of Glasgow, not failed, this steam frigate, known first as the *Victoria* and then as the *British Queen*, might have become the first of all transatlantic steamships. But the company which Smith had formed, and of which Macgregor Laird, the African traveller, had become the secretary, was not to be allowed to have matters all its own way, for while its directors were busy in London, another company was formed at Bristol under the name of the Great Western Steamship Company, and a third at Liverpool called the Transatlantic Steam Ship Company. To the London company we owe the voyages of the *Sirius*, the *British Queen*, and the *President*; to the Bristol company those of the *Great Western*; while the ships despatched by the Liverpool company were the *Royal William* and the *Liverpool*. Just as Smith and Macgregor Laird were the most active members of the London company, so Isambard Kingdom Brunel (1806–59) was the principal figure in the Bristol company, while the Liverpool company owed much to Charles Wye Williams (1780–1866), a lawyer who had turned his attention to steam navigation.

To follow the fortunes of these three pioneer companies and of the Cunard Company which came on the scene a little later, would require a volume, and all that can be done here is to give some particulars of the vessels they either chartered or owned. Many of the principal facts connected with them are to be found in such works as Lindsay's *History of Merchant Shipping*, Maginnis's *Atlantic Ferry*, and Preble's *History of Steam Navigation*. From none of these sources, however, can a complete account of the events of the years 1838 to 1840 be obtained, but further interesting matter is to be found in a small MS. volume entitled *Glances at Atlantic Steam Navigation, 1838–1841*, included in the Field papers in the Science Museum Library. The notes contained in this volume were evidently compiled by a member of the firm of Maudslay, Sons and Field while the vessels were running, and are therefore of unique value.

For the sake of conciseness, Tables I and II have been drawn up giving the principal dimensions of the six vessels which ran to New York and also of the *Britannia*, the first of the Cunard ships which ran to Halifax. These are followed by Table III giving the dates of their maiden voyages, as well as those of the *Acadia* and *Caledonia*, sister ships of the *Britannia*, and the number of voyages made by each vessel up till the end of 1840. The ships were all of wood, while the engines were of the side-lever type.

From these tables it will be seen that the *Sirius* and the *Royal William*—the latter not to be confused with the *Royal William* which crossed from Canada in 1833—were much smaller than the others. Neither of them

## Table I. Particulars of ships

| Ship | Length ft. | Breadth ft. | Tonnage | Builder |
|---|---|---|---|---|
| *Sirius* | 178 | 25½ | 703 | Menzies |
| *Great Western* | 236 | 35¼ | 1320 | Patterson |
| *Royal William* | 175 | 27 | 617 | Wilson |
| *Liverpool* | 223 | 30⅝ | 1150 | Humble and Milcrest |
| *British Queen* | 245 | 40½ | 1863 | Curling and Young |
| *Britannia* | 207 | 34¼ | 1154 | Duncan |
| *President* | 243 | 41 | 2366 | Curling and Young |

## Table II. Particulars of engines

| Ship | Diameter of cylinder in. | Stroke ft. in. | Horse-power | Maker |
|---|---|---|---|---|
| *Sirius* | 60 | 6  0 | 320 | Wingate |
| *Great Western* | 73½ | 7  0 | 400–750 | Maudslay |
| *Royal William* | — | — | 270 | Fawcett and Preston |
| *Liverpool* | 75 | 7  0 | 468 | Forrester |
| *British Queen* | 77½ | 7  0 | 500 | Napier |
| *Britannia* | 72½ | 6  10 | 440–740 | Napier |
| *President* | 80 | 7  6 | 540 | Fawcett and Preston |

## Table III. Voyages, 1838–40

| Ship | Port | Date of departure | Port | Date of arrival | No. of voyages 1838–40 |
|---|---|---|---|---|---|
| *Sirius* | Cork | 4. 4.38 | New York | 23. 4.38 | 2 |
| *Great Western* | Bristol | 8. 4.38 | ,, | 23. 4.38 | 17 |
| *Royal William* | Liverpool | 5. 7.38 | ,, | 24. 7.38 | 3 |
| *Liverpool* | ,, | 22.10.38 | ,, | 23.11.38 | 7 |
| *British Queen* | Portsmouth | 12. 7.39 | ,, | 27. 7.39 | 8 |
| *Britannia* | Liverpool | 4. 7.40 | Boston | 18. 7.40 | 3 |
| *President* | ,, | 1. 8.40 | New York | 17. 8.40 | 2 |
| *Acadia* | ,, | 4. 8.40 | Boston | — | 2 |
| *Caledonia* | ,, | 1. 9.40 | ,, | — | 2 out, 1 home |

was intended for the Atlantic passage. At the beginning of 1838, however, as the *Great Western* was nearing completion, while the *British Queen* was far from ready, the London company chartered the *Sirius* from the St George's Steam Packet Company, and despatched her from Cork four days before the much larger *Great Western* was due to leave Bristol, and the premier honours thus fell to the *Sirius*. Her first voyage out was not accomplished without considerable difficulty, there being little coal in her bunkers when she reached New York, and though, after returning home in May, she made another passage out and home, she proved uneconomical and was returned to her original owners. Abstracts from the log of the *Sirius* were published at the time, and among the remarks are such entries as "Pressure on boilers $5\frac{3}{4}$ lb." "Mixed resin with the picked ashes;" "one ton of coals burned one hour and fifteen minutes, without ashes or resin" and "Fresh water in the boilers all the way, with Hall's condensers". It was not a little remarkable that the *Sirius* and also the *British Queen* were both fitted with Samuel Hall's surface condensers, a great novelty at that time.

The *Great Western* was undoubtedly the most remarkable of the pioneer ships on the Atlantic, and reflected the greatest credit on all who had assisted in her construction. Designed by Brunel and built by Patterson, she was launched on July 19, 1837. Her scantling, wrote the compiler of the Field manuscript,

is equal in size to that of our line of battleships, it is filled in solid, caulked within and without up to the first futtock heads before planking, and all above this height is of English oak. The flooring is of immense length, each floor overrunning the other; they are strongly dowelled and bolted in pairs first and also together by 1½-in. bars of 24 ft. in length, driven in parallel rows, scarfing about 4 ft. She is trussed in the firmest and closest manner by iron and wooden diagonal and also shelf pieces, these, including all her upper works, are fastened with screws and nuts to an immense extent.

Her machinery by Maudslays was worthy of the ship. She had four flue boilers 11½ ft. long, 9½ ft. wide and 16¾ ft. high, the flues being arranged above the furnaces. Bourne, in his *Treatise on the Steam Engine*, 1846, gives drawings of these boilers, which were worked at 5-lb. pressure, and of the tubular boilers by which they were replaced. The side-lever engines, weighing about 200 tons, were bolted to "a vast frame, the beams of which extend to and press equally on every part of her bottom". After steaming some 75,000 miles she was inspected in dry dock by a Lloyd's Register Surveyor, who reported that he "found her to be in perfectly good condition, free from any indication of defect inside: the copper smooth and without any apparent strain outside". The first voyage of the *Great Western* lasted from April 8, 1838, to April 23, her average day's run being 215 miles. Altogether, she made five double passages in 1838, six in 1839, and before she was disposed of in 1846 crossed and recrossed seventy-four times. No other ship had a finer record and none attracted more attention. While she was still running on the Atlantic, Sir John Rennie, referring to her first passage, remarked that "the success of this voyage across the Atlantic having exceeded the most sanguine expectations of its promoters and, indeed, of the world, there seemed no bounds to the extension of steam navigation."

The year 1838 also saw the first voyages of the *Royal*

*William* and the *Liverpool*. The Transatlantic Steam Ship Company had purchased the *Liverpool* while still on the stocks, but to start their service they chartered the *Royal William* from the City of Dublin Steam Packet Company. Of only 617 tons and 270 horse-power, the *Royal William*, like the *Sirius*, proved unsuitable for the trade, and after three voyages she ceased to run. But small as she was, she was easily able to beat the sailing packets, on one occasion overhauling the *Hibernia* which had left Liverpool eighteen days before her, and the *Sir James Kempt*, which was already fifty-eight days out from Dundee. On her first Atlantic voyage she carried a certain amount of compressed peat, of which the captain stated, "1 cwt. of this saves 3 cwt. of coal"! The *Liverpool* was a far finer vessel than the *Royal William* and was the largest constructed on the banks of the Mersey up till that time. She left Liverpool October 20, 1838, was driven back to Cork by stress of weather, sailed again a few days later and reached New York November 23. Her best day's run was 242 miles, her worst 151. She returned home in December, and in 1839 made six round voyages. On her last return voyage of that year she was driven as far south as the Azores, had to refill her bunkers and was twenty-seven days on passage. Taken in hand for alteration, she was destined not to sail the Atlantic again, for in July 1840 the Transatlantic Steam Ship Company was wound up, and the *Liverpool* became the property of the Peninsular and Oriental Steam Navigation Company.

The fifth ship to be added to the Atlantic service was the *British Queen*. Described as "a most beautiful specimen of London shipbuilding, and for elegance of mould, great strength and admirable proportion of parts, thought by many to be unequalled", this vessel, for which the contract had been placed in 1836, was not ready for launching till May 1838, and her completion was much delayed by the failure of the engine makers. This failure necessitated another contract being made, and both engines and boilers were finally made by Robert Napier. Sketches of her boilers are given by Bourne. She sailed from Portsmouth on her maiden voyage on July 12, 1839, and reached New York on July 27. On her return journey she left New York on August 1, the same day as the *Great Western*, the latter arriving at Bristol on August 14 and the *British Queen* arriving at Portsmouth the following day. She made two more voyages in 1839, and in 1840 crossed and re-crossed five times, but she proved considerably slower than Brunel's fine ship. Her first voyage outward in 1841 began on March 10, and, running into the gales which swept the Atlantic that month, and in which the *President* foundered, for 220 hours the *British Queen* was buffeted by tremendous seas. She had serious trouble with her paddle floats, her sails were torn to shreds, and she arrived in New York after a passage of twenty-four days. She left for home again on April 11, reached Liverpool on April 28, and was soon afterwards sold to a Belgian company. The *British Queen* was on her third voyage of 1839 when her sister ship, the ill-fated *President*, was launched at Limehouse. A somewhat shorter but broader vessel than the *British Queen*, the *President* was the largest of all the pioneer steam ships on the Atlantic. Though well built with timbers of oak diagonally fastened fore and aft with iron, and presenting a "very handsome appearance to the superficial beholder", by many she was thought to be ill-adapted for fast sailing. Unfortunate from the first, she was nearly lost in the Channel when being towed to Liverpool to have her machinery fitted. Her first voyage to New York took 16½ days. Her second was made in October and November 1840. On February 10, 1841, she left Liverpool on her third voyage and arrived at New York on March 3. Eight days later, in command of Lieutenant Richard Roberts, who had been captain of both the *Sirius* and *British Queen*, she left New York for home, and at once met with bad weather. She was last seen by the sailing vessel *Orpheus* at 4 p.m. on March 12, between Nantucket Shoal and St George's Bank, labouring heavily and shipping much water. Nothing more was heard of her. The gale continued with unabated fury till midnight of March 13, but it was thought the *President* foundered in the night of the 12th. With her loss and the sale of the *British Queen*, the activities of the company founded by Junius Smith came to an end.

The founder of the British and North American Royal Mail Steam Packet Company, for which the *Britannia*, *Acadia*, *Columbia* and *Caledonia* were constructed, was Samuel Cunard (1787–1865), a native of Nova Scotia. But whereas the financial prosperity of this company owed much to the work of Sir George Burns (1795–1890) and David MacIver, the success of the ships themselves was entirely due to the technical knowledge and ability of Napier, who at the time the company was formed was established in Glasgow as a marine engineer. From the *Life of Robert Napier* it will be seen that Cunard's first suggestion was for ships of only 800 tons and 300 nominal horse-power. Napier concluded that ships of that size were not likely to be suitable for Atlantic traffic and it was on his advice that larger and more powerful ships were built. As will be seen from the tables, the *Britannia*, the first Cunard ship to sail, was of 1154 tons with engines of 740 indicated horse-power. The engines had two cylinders $72\frac{1}{2}$ in. diameter with a stroke of 6 ft. 10 in., and they were fitted with expansion valves. At 16 revolutions per minute the engines gave the ship a speed of about $8\frac{1}{2}$ knots, while the coal consumption was from 31 to 38 tons per day.

When the *Britannia* left Liverpool on her maiden voyage to Halifax and Boston, her commanding officer, Captain Woodruff, was furnished with a series of rules relating to the sailing and steaming of the ship. Among these rules were the following:

You should make it the duty of the officer on deck to ask the engineers at regular intervals if they have blown off the boilers, and count the revolutions of the engines every two hours, putting them down in the log. Your consumption will be light if the furnace doors are kept as much shut as possible, and a judicious use made of the dampers. As she lightens work her expansively; if you have smart leading wind 2nd or 3rd grade should carry her fast long. The first engineer should be allowed to have the general arrangement of all connected with the engine room. Some of the duties we expect from him are as follows: To weigh coal 4 times a day. Keep a steam log. Keep an accurate account of the stores he uses about the engine—oil, tallow, etc. Write into a book or books, daily, the coal consumed, the steam log, and the expenditure of engine room stores. The general duties of the 5 engineers are the same—great care of all connected with the engines and boilers. We have confidence in their giving every attention to this. Burn up all the ashes; throw nothing overboard. There are riddles on board.

The *Britannia* was the first of the Atlantic steamships to be entrusted with the mails and the Cunard Company was the first to be given a subsidy in connection with this duty. The *Great Western*, though admirably suited to the purpose, was never allowed to carry mails, but her performance gave rise to the humorous volume, *The Letter Bag of the Great Western* by Sam Slick, otherwise Judge Thomas Chandler Haliburton (1796–1865), who in the last of the series of letters the book contains said "Since the discovery of America by Columbus nothing has occurred of so much importance to the New World as navigating the Atlantic by steamers".

Table IV

|  | Longest outward | Shortest outward | Average outward | Longest home | Shortest home | Average home |
|---|---|---|---|---|---|---|
| Sailing ships | Days Hours | Days Hours | Days Hours | Days Hours | Days Hours | Days Hours |
| Old or Black Ball Line | 48 | 22 | 33–17 | 36 | 18 | 22–12 |
| Dramatic Line | 38 | 23 | 30–12 | 25 | 17 | 20–12 |
| Star Line | 45 | 27 | 36 | 28 | 21 | 24 |
| Swallow Tail Line | 45 | 28 | 35 | 31 | 17 | 22–12 |
| Steam ships |  |  |  |  |  |  |
| *Great Western* | 21–12 | 13 | 16–12 | 15 | 12–6 | 13–9 |
| *British Queen* | 20–9 | 14–21 | 17–8 | 21–12 | 13–12 | 16–12 |
| *Liverpool* | 18–12 | 16 | 17–4 | 27 | 13–18 | 15–16 |

There are many other matters connected with the grand achievement of transatlantic steam navigation which deserve consideration but with which it is not possible to deal here. Attention however may be drawn to some of the results. Steam introduced into Atlantic travel a certainty and regularity quite unknown previously, while there was a reduction in the time spent on passage of approximately one-half. In the *Shipping Gazette* and the *Civil Engineers' Journal* for 1840 information was given regarding the time taken by the best sailing packets and the three steam ships *Great Western*, *British Queen* and *Liverpool*, from the sailing of the *Great Western* on April 8, 1838, to the arrival home of the

*Liverpool* on January 11, 1840. From these sources Table IV has been prepared.

From this table it will be seen that the use of steam effected a saving of time of about seventeen days on the outward voyages and about seven days on the homeward voyages.

## CHAPTER IV

## STEAM MEN-OF-WAR

THE years which saw the building of the first Atlantic steamships also saw a notable advance in the number, size and power of our paddle-wheel warships, and by the middle of last century the Navy possessed a fleet of steam frigates and steam sloops which had done good work in all parts of the world, and had effectually demonstrated their value in naval operations. The increase in the number of steam vessels had led to the formation of a Steam Department at the Admiralty in 1837, and to the organisation of the Engineering Branch afloat the same year. It had likewise opened up new possibilities for naval officers, of which many lieutenants and young commanders were not slow to take advantage. In the command of the paddle vessels they saw chances of advancement at a time when promotion was notoriously slow, and to some of these officers we are indebted for interesting accounts of the vessels they commanded. The history of the paddle-wheel warships moreover recalls many early experiments in marine engineering, and introduces us to the historic firms of Maudslay, Penn, Seaward, Miller and Rennie, whose work made the Thames a great marine engineering centre. At first, Boulton, Watt and Co. and the Butterley Iron Works obtained most of the naval contracts, but of the later and larger paddle-wheel vessels out of a total of seventy, no fewer than fifty were engined by the Thames firms, and the succeeding pages have been written partly with the object of directing attention to their early history.

Particulars have already been given of some of the earlier steam vessels in the Navy, but it is worth recalling that the first attempt to drive a naval vessel by steam was made by the versatile inventor, Charles, third Earl Stanhope (1753–1816). According to Philip, fifth Earl Stanhope (1805–75), who mentions the matter in his *Life of Pitt*, the Navy Board had a ship built on the Thames expressly for his experiments, and in 1794 Stanhope entered into a bond of £9000 "to indemnify the public in case the said ship should not answer the purpose of Government". "The Ambi-Navigator Ship, called the *Kent*" was moored in Greenland Dock while being fitted out, and Stanhope declared it his intention to establish every part of the subject on clear and irrefragable proofs and to ascertain demonstratively what was the best possible plan. Progress not being as rapid as he anticipated, in December 1795 he applied for and obtained an extension of twelve months for the experiments, which he declared were on the eve of conclusion. Finally, however, the matter was dropped, for as Robertson Buchanan, the writer of the first book on marine engineering, *A Practical Treatise on Propelling Vessels by Steam*, who had seen the engine of the *Kent*, tells us, the mechanism did not answer his Lordship's expectations.

Twenty years passed before the Admiralty again acquiesced in money being spent on a steam vessel, and the second scheme was no more successful than the first. In 1815, Captain J. H. Tuckey, R.N., was fitting out a vessel called the *Congo* for exploring the river of that name. On the suggestion of Sir Joseph Banks, then the acknowledged head of British science, it was decided to convert the *Congo* to a steamer, to enable her to stem the current in the river. With this end in view, Sir

Robert Seppings, the Surveyor of the Navy, obtained an overhead beam engine from Boulton, Watt and Co. which, however, by its weight reduced the freeboard of the vessel to a dangerous extent, and at its best propelled her at only four knots. The plan was thereupon

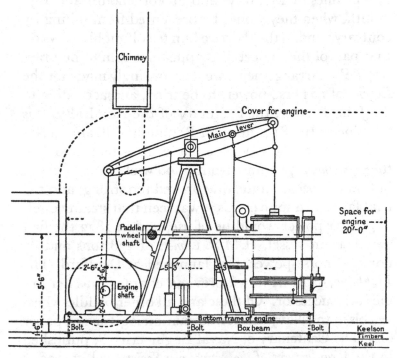

Fig. 1. Paddle-wheel engine for *Congo*, 1815
Boulton and Watt Collection. Courtesy of Birmingham Reference Library

abandoned and the engine, the drawings of which are preserved in the Birmingham Reference Library, was afterwards installed as a stationary engine in Chatham Dockyard. One of these drawings is reproduced in Fig. 1.

Soon after this it was impressed upon the Admiralty by John Rennie and the elder Brunel, that steam vessels would be of use at the various naval ports, and to Brunel's representations Lord Melville replied that as the application of the steam engine "would be attended with material advantage to his Majesty's Service if it could be used for towing ships of war out of harbour, in the Thames or Medway, and at Portsmouth and Plymouth, when they would be prevented from sailing by contrary winds, [they] desire him to submit his views on that part of the subject, if it appears to him to be practicable". Arrangements were accordingly made for the *Regent*, of 16 horse-power, to be tried, and according to Sir John Barrow, the Secretary of the Admiralty, this was done. In 1819, through Rennie, the steam packet *Eclipse*, of 60 horse-power, made an attempt to tow the *Hastings*, 74 guns, from Woolwich to Chatham, but on this occasion the tide proved too strong, and the trial failed. In spite of this it was seen that steam vessels of greater power would be of use, and in 1822 the first naval steam vessel, H.M.S. *Comet*, of 238 tons and 80 nominal horse-power, was launched at Deptford. The *Lightning*, dealt with in a previous chapter, was her immediate successor. Of these and all other paddle-wheel vessels in the Navy much information is given in a paper on "The Centenary of Naval Engineering", published in the *Transactions of the Newcomen Society*, vol. II, 1921–22; while in *Notes and Queries* for May 1927 will be found a carefully compiled "Chronological Order of the Introduction of Steam-Propelled Vessels into the Royal Navy", by Commander J. A. Rupert-Jones, R.N.R.

By 1837 there were 27 steam vessels included in the Navy List, and these were classified by Commander Otway in his *Treatise on Steam* in the following groups:

| Group 1 | Group 3 | Group 5 |
|---------|---------|---------|
| *Cyclops* | *Hermes* | *Lightning* |
| *Gorgon* | *Firebrand* | *Meteor* |
| | *Firefly* | *Confiance* |
| Group 2 | *Megaera* | *Echo* |
| *Dee* | *Spitfire* | *Alban* |
| *Medea* | *Volcano* | *Carron* |
| *Rhadamanthus* | *Flamer* | *African* |
| *Phœnix* | | *Comet* |
| *Salamander* | Group 4 | |
| *Messenger* | *Blazer* | |
| | *Tartarus* | |
| | *Columbia* | |
| | *Pluto* | |

The first two vessels in this list, the *Cyclops* and *Gorgon*, were then building at Pembroke. They were the first steam vessels in the service of over 1000 tons measurement, and were regarded as the first steam frigates. In the succeeding thirteen or fourteen years the numbers increased rapidly. Particulars of eight of the largest of

Table V

| Year | Ship | Place built | Tons (B.O.M.) | N.H.P. | Maker of engines |
|------|------|-------------|---------------|--------|------------------|
| 1844 | *Retribution* | Chatham | 1641 | 800 | Maudslay |
| 1845 | *Terrible* | Deptford | 1847 | 800 | Maudslay |
| 1846 | *Odin* | Deptford | 1326 | 560 | Fairbairn |
| 1846 | *Sidon* | Deptford | 1328 | 560 | Seaward |
| 1849 | *Magicienne* | Pembroke | 1255 | 400 | Penn |
| 1850 | *Leopard* | Deptford | 1435 | 560 | Seaward |
| 1850 | *Furious* | Portsmouth | 1286 | 400 | Miller |
| 1851 | *Valorous* | Pembroke | 1258 | 400 | Penn |

the ships are given in Table V, the *Valorous* being the last paddle-wheel frigate to be added to the Navy. Though a few iron vessels had been constructed, all these eight were of wood. The tonnage is that given by the so-called "builders' old measurement", or

"B.O.M.", while the horse-power is that given in the Navy Lists. The indicated horse-power was from twice to three times the nominal horse-power.

Though even such small vessels as the *Comet* and *Lightning* carried guns, the first steam vessels built as fighting ships were the *Dee, Medea, Rhadamanthus, Phœnix* and *Salamander*, included in Otway's Group 2. All these were fitted with side-lever engines by the firm of Maudslay, Sons and Field. This firm had been founded by Henry Maudslay (1771–1831) in 1797, but it was the work of his sons, Thomas Henry Maudslay (1792–1864) and Joseph Maudslay (1801–61), and of their partner, Joshua Field (1787–1863), which made the shops at Lambeth so famous. Of the five vessels mentioned above, for which they constructed the engines, the *Medea* was regarded as the most successful, and in a *Memoir of Her Majesty's Steam Ship The Medea during a Service of nearly Four Years*, by Thomas Baldock, Lieut. R.N., we have an excellent account of her first commission. The *Medea* was 179 ft. 4½ in. between perpendiculars, 46 ft. wide over the paddle-boxes, and 835 tons by the old measurement rules. Her engines were of 220 nominal horse-power, the weight of the engines being 165 tons, that of the boilers 35 tons, and of the water in the boilers 45 tons. With 320 tons of coal, 18 tons of water and three months' provisions aboard, she drew 13 ft. 10 in. forward and 14 ft. 6 in. aft. According to Baldock, she could, altogether, steam 1190 miles at 8½ knots, 1258 miles at 9 knots, and 1360 miles at 10 knots, and was considered fit for any passage of 3000 miles save a westerly voyage across the Atlantic in winter.

Launched in September 1833, the *Medea* was soon commissioned, and two incidents, recorded by Baldock, illustrate one of the uses to which she was put. On January 18, 1835, the fleet of line-of-battleships had arrived about 10 miles off Malta when it became becalmed. Orders were then given to the *Medea* to raise steam, and she proceeded to tow each vessel in turn into harbour. About a fortnight later it was desired to get the fleet to sea, but heavy weather prevented the ships from sailing. Again the services of the *Medea* were requisitioned, and on February 8, with the ramparts of Valetta crowded with spectators, she proceeded to tow the *Caledonia*, of 120 guns, and five 80-gun ships clear of the harbour.

In this manner [said Baldock] all the fleet were taken out, the whole being effected in four hours and ten minutes, after which the fires were extinguished on board the *Medea*, and she pursued her course as a sailing vessel in company with the squadron; the whole cost of coal and other engine stores expended on this occasion amounting only to 3*l.* 6*s.*

That the machinery was well constructed and equal to its task may be presumed, for during the whole commission, which lasted from February 1834 to October 1837, no repairs were done by any one but the ship's staff. Thomas Hamshaw, afterwards long known as the Engineer of Malta Dockyard, was in charge of the machinery, and of him his commanding officer, Commander Austin, wrote that he was "as competent an engineer as was to be met with afloat".

The *Medea* was practically the last vessel engined by Maudslays with side-lever engines, for in 1839 Joseph Maudslay and Field patented the twin-cylinder engine, and this was the type supplied to the four large paddle-vessels, *Scourge*, *Devastation*, *Retribution*, and *Terrible*.

Brought out with the object of obtaining a long-stroke engine in the height available, engines of this type had two pairs of cylinders, each pair having a common connecting rod. As just about the time these engines were being fitted the famous Siamese twins, Chang and Eng, were being exhibited in the country, the Maudslay engine was nicknamed the "Siamese" engine, and it is so called in the text-books. Bourne gives a drawing of the engines of the *Retribution*, while in the Science Museum is a model of the engines of the *Devastation*. This later model was exhibited, with others, at the Great Exhibition of 1851, when it was stated that the firm had constructed marine engines of various types with an aggregate of 35,183 horse-power. The arrangement of a Siamese engine can be seen from Fig. 4, p. 74.

The finest set of Siamese engines were those of H.M.S. *Terrible*, the steam frigate constructed in 1845. The engines of the *Terrible* had four cylinders, 72 in. diameter and 8-ft. stroke. Steam at 15 lb. per square inch was supplied from tubular boilers, and the engines of 800 nominal horse-power, at 16 revolutions per minute, developed 2059 indicated horse-power. To withstand the stresses due to so powerful an engine, the *Terrible*, like the *Great Western*, was built with her timbers close together forming a watertight body before the external planking was added. She was 226 ft. between perpendiculars, 1847 tons by measurement, and 3189 tons displacement, and had been designed by Oliver Lang, Master Shipwright, who had also designed the *Medea*. Soon after commissioning, the *Terrible* joined the flag of Admiral Sir William Parker, who wrote from the Mediterranean: "The *Terrible* is a noble ship and answers completely." With the *Sampson*, *Retribution*, *Tiger*, *Furious*, and three French steam vessels, *Mogador*, *Vauban* and *Descartes*, the *Terrible* took part in the bombardment of Odessa on April 22, 1854, and she successfully rode out the great gale of November 14, 1854, which wrought great havoc in the Black Sea Fleet. In later years she was employed in the Atlantic cable-laying operations, and helped to steer the floating dock to Bermuda.

Maudslays was the only Thames firm to execute naval contracts until about 1835, when orders were given to Miller and Barnes of Ratcliff, and Seaward and Capel of Millwall, whose connection with the Navy began with the construction of the engines of the *Volcano*. John Seaward (1786–1858) was the founder of the Millwall firm, and he was long known as one of the most energetic and progressive engineers of his time. The son of a Lambeth builder, he gained experience in bridge building, lead mining and gas engineering, and in 1824, with his brother, opened the Canal Ironworks, Millwall. He was an advocate of tubular boilers and surface condensing, and he invented a telescopic funnel, brine valves for boilers and a disconnecting crank. In 1836, with the object of saving weight and space, he brought out his direct-acting paddle engine, with the cylinder directly beneath the crank and with a short connecting rod. The piston rod end was guided by a parallel motion, and as the first set of this type of engine was fitted in the *Gorgon*, the engine became known as the "Gorgon" engine, and that name was also applied to engines of other makers having somewhat similar features. A model of the *Gorgon's* engine is in the Science Museum; a diagram of the engine is given in Fig. 2. The *Gorgon* was a notable vessel in many ways. Designed by Captain Sir William Symonds, who from 1832 to 1847 was responsible for the design of

most of the paddle-wheel vessels, she was built of teak
and was 178 ft. between perpendiculars, of 37½ ft.
beam and 1108 tons B.O.M. Her sister ship, the
*Cyclops*, was slightly larger, being 1195 tons B.O.M. and
1862 tons displacement. The engines had two cylinders,
64 in. diameter, and 5½ ft. stroke, and another innovation was the use of separate steam and exhaust valves at
each end of the cylinders. The cylinders were carried
on large bed plates forming the condenser and hot-well,

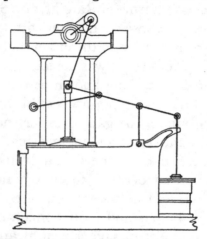

Fig. 2. Diagram of Seaward's engine for *Gorgon*, 1837

while the entablature was carried on wrought-iron
pillars. The nominal horse-power of the engine was 320,
and the machinery, including the boiler water, weighed
276 tons. Spoken of as a very superior vessel, fast and
particularly easy and dry, and able both to sail and
steam well, the *Gorgon* took part in the operations off
the Syrian Coast in 1840, when Admiral Stopford wrote
to the Admiralty that "the steam vessels have been
eminently useful in constantly moving along a great
extent of coast with troops and arms and taking part in
the attack upon the different forts." A year or two

later the *Gorgon* was sent to South America where she was nearly lost. On May 10, 1844, she was driven ashore in the Bay of Monte Video and sank 10 or 12ft. in the sand. On board was a young lieutenant, Cooper Key, destined to become afterwards First Sea Lord. Key was one of the officers who deliberately took up the study of steam, and when exchanging into the *Gorgon* from a sailing vessel he looked upon the step almost as a change of profession. To him we owe *A Narrative of the Recovery of H.M.S. Gorgon*, giving an account of the salvage operations. Opinion had held that in the case of a steamer stranding, the engines should be taken out, but the captain of the *Gorgon*, decided not only to keep them on board but to utilise them to work a bucket dredger to help free the ship. This apparatus, however, was never put into action, but the engine was used for hauling the vessel back into the sea. After the *Gorgon*, Seaward supplied direct-acting engines to the *Cyclops*, *Alecto*, *Polyphemus*, *Driver*, *Caradoc* and *Leopard*, his largest engines being those of 560 nominal horse-power fitted in the *Sidon*, the paddle vessel designed by Captain Charles Napier.

The firm of Miller and Barnes had been started at Glasshouse Fields, Ratcliff, in 1822, by Joseph Miller (1797–1860) and John Barnes (1798–1852). The latter was a godson of Watt, and both he and Miller received their early training under Murdock at the Soho Foundry. Barnes left the firm in 1835, afterwards doing notable work in France, but under the various names of Miller and Ravenhill, and Ravenhill and Salkeld, the firm had a long connection with the Navy. Following on the lines of Seaward and the Maudslays, Miller also brought out a direct-acting paddle-wheel engine, and fitted engines of this type to the *Rosamund, Gladiator* and *Barra-*

*couta*, and also supplied oscillating engines to other vessels. A man of artistic and literary tastes, Miller, in his engines, endeavoured to combine strength with lightness, and they were always known for their graceful proportions. Miller retired from the firm in 1852 and died in South Carolina, where he had settled.

The fourth Thames firm to receive contracts for naval machinery was that of John Penn and Sons. Founded at Greenwich in 1800 by the Somerset millwright, John Penn (1770–1843), it was the second John Penn (1805–78) who made it the rival of Maudslays. Marine engine building was taken up at Greenwich in 1825, and it was there the oscillating engine was brought to its greatest degree of perfection. An account of the steps leading to the long association of the firm with the Navy is given by Sir John Rennie. The Admiralty, in 1843, had for their official yacht the *Black Eagle*, originally named the *Firebrand*. Desiring to improve her speed they consulted Boulton, Watt and Co. on the possibility of fitting more powerful engines, but received a discouraging reply. Penn, hearing of this, uninvited sent a tender offering to fit new machinery to the *Black Eagle*, of the same weight and occupying the same space as her original engines but of double the horse-power. On the receipt of this tender the Comptroller, Admiral Sir Byam Martin, asked Sir John Rennie's advice, on hearing which the offer was accepted and Penn satisfactorily fulfilled his promise. The *Black Eagle* was a small vessel which, after lengthening, was only 540 tons but had engines of 260 nominal horse-power. The engines were of the oscillating type and Penn supplied similar engines to the *Sphynx*, *Magicienne*, *Argus*, *Valorous*, and *Banshee*.

The speed of the paddle-wheel frigates was about

10 knots, but in the *Banshee*, which was built with three
others for the Dublin and Holyhead traffic, a speed of
16 knots was attained. In speaking of the *Banshee* and
her sister vessels, Fincham, in his *History of Naval Architecture*, says their building was the most successful effort
on the part of the Admiralty at producing fast steamers.
Of the four, the *Banshee* was the fastest, and in Table VI
are given some of the particulars from Fincham's
*History*.

Table VI

| | Caradoc | Banshee | Llewellyn | St Columba |
|---|---|---|---|---|
| Length, b.p. (ft. in.) | 193 0 | 189 0 | 190 0 | 198 6 |
| Beam (ft. in.) | 26 9 | 27 2 | 26 6 | 27 3 |
| Burthen (tons) | 662 | 670 | 650 | 719 |
| Engine maker | Seaward | Penn | Miller | Laird |
| Cylinder, diam. (ft. in.) | 6 2 | 6 0 | 5 8 | 5 10 |
| Stroke (ft. in.) | 6 0 | 5 6 | 4 4 | 5 6 |
| Pressure per sq. in. (lb.) | 14 | 14 | 20 | 14 |
| Nominal horse-power | 350 | 350 | 350 | 350 |
| Speed of ship (knots) | 14 | 16·32 | 15·2 | 14·23 |

One other Thames firm, that of J. and G. Rennie,
was associated with the Navy at this time. Founded in
1821 by the younger John Rennie (1794–1874), who
was knighted in 1831, and his brother George Rennie
(1791–1866) the firm carried on engine making in the
old works established by the elder Rennie in Holland
Street, Blackfriars in 1791, and in after years executed
many important contracts. They constructed the
machinery for the little *Mermaid* of 164 tons which as
H.M.S. *Dwarf* became the first screw vessel in the
Navy, and in the middle 'forties built paddle-wheel
engines for the *Bulldog*, the iron vessel *Oberon* and the
*Sampson*. It was in the last of these they fitted the shifting link motion usually called the Stephenson link

motion, instead of the then usual single loose eccentric with stops on the shafts.

To these Thames firms was afterwards added that of Humphrys, Tennant and Dykes, founded at Deptford in 1852, mainly through the efforts of Edward Humphrys (1808–67), who had worked with Penn and the Rennies, and had held the post of Chief Engineer of Woolwich Dockyard. Established too late to secure contracts for paddle-wheel engines, in the early 'fifties the firm built the engines for several screw sloops and thus began a connection of half-a-century with the Navy, during which it was the rival of Maudslays and Penns.

All these early marine engineers were well known in engineering and scientific circles, and they served the interests of their profession in many ways. Two members of the Maudslay family were among the founders of the Institution of Civil Engineers, which Sir John Rennie and Field both served as president. The Miller medal was founded out of funds left to the Institution by Joseph Miller. Miller too, like Field, Penn and George Rennie, was a fellow of the Royal Society, while Penn was twice president of the Institution of Mechanical Engineers. Though the works they established have long since disappeared, during the nineteenth century engines were built in them for practically every naval and mercantile fleet in the world. Penn, up till the time of his death, had alone fitted 735 vessels with engines, while the long list of ships engined by Maudslays included the famous White Star liners *Oceanic*, *Germanic* and *Britannic*, the two fast cruisers *Iris* and *Mercury* and the *Blake* of 20,000 horse-power. The last naval contracts executed by Maudslays, whose connection with the Navy dated from 1823, were those for the machinery of the battleships *Albion*, 1898, and *Venerable*, 1899; Humphrys engined the battle-cruiser *Invincible* in 1907, while from the works of Penn in 1911 came the machinery for the cruiser *Chatham* and the battleship *Thunderer*, with the completion of which the marine engineering industry on the Thames may be said to have come to an end. Many factors had contributed both to its rise and to its decline, but the student of its history cannot have anything but admiration for the lives, characters and achievements of its pioneers.

## CHAPTER V

# THE INTRODUCTION OF SCREW PROPULSION

SIMULTANEOUSLY with the inauguration of transatlantic steam navigation and the construction of paddle-wheel fighting fleets, to which the two preceding chapters have been devoted, came the introduction of the screw propeller, the greatest of all innovations in the propulsion of ships by mechanical power. Though generally associated with the name of the great geometer, Archimedes, who lived in the third century B.C., we are told that the screw was probably used for various purposes even before his time. But in spite of the antiquity of the screw, it remained for modern inventors to adapt it for driving vessels.

Of such inventors, however, there have been many. Bourne, in his work on the screw propeller, in 1867, enumerates some 470 names associated with the invention. It may be recalled that Watt, when writing to his friend Dr Small, in 1770, on the subject of propelling canal boats, asked, "Have you ever considered a *spiral oar* for that purpose?" Mathematicians, like Daniel Bernouilli and Paucton, wrote on the screw, but the earliest experiments date from the beginning of the nineteenth century, and the first important trials from the 'thirties. Critics and pessimists were not slow in foretelling failure, and one writer, speaking of the screw propeller, said "the whole engineering corps of the empire was arrayed in opposition to it, alleging that it was constructed on erroneous principles and full of practical defects". The pioneers assuredly had many difficult

PLATE III

ROBERT NAPIER
(1791–1876)
Courtesy of James Napier

JOHN ERICSSON
(1803–89)
Courtesy of Science Museum

ISAMBARD KINGDOM BRUNEL
(1806–59)
Courtesy of National Portrait Gallery

SIR FRANCIS PETTIT SMITH
(1808–74)
Courtesy of Science Museum

problems to solve, for not only had they to discover the best form of screw and where it should be placed, but they had to devise means of driving it fast enough. They had to find out how to take the thrust, how to construct the under-water stern bearing, how to lift or feather the screw, and how to diminish as far as possible the vibration which proved so injurious to wooden ships.

The first British ship ever moved by a screw was the Government transport *Doncaster*. In June 1802 the *Doncaster* left England for Gibraltar, having on board a shaft and screw propeller and some rope pulleys. The propeller was the "perpetual sculling machine" of Edward Shorter, of London. Arrived at Gibraltar, the necessary framing and pulleys were placed out-board, the propeller was shipped, the ropes were rove, and with eight men walking the capstan around, the *Doncaster* was moved up and down Gibraltar Bay at about $1\frac{1}{2}$ knots. The feat attracted considerable attention among the naval officers at Gibraltar and again at Malta, where the "perpetual sculling machine" enabled the ship to enter the harbour in a calm. The trials at Malta were witnessed by Admiral Sir R. Bickerton, whose certificate, with that of Captains Aylmer and Keats, is preserved in the Kelvingrove Museum, Glasgow. "Having been on board the *Doncaster* transport", wrote Admiral Bickerton to John Shout, the master, "and examined the working of the propeller while the ship was under way, I have to inform you that I think the plan a good one, and that it may, in many instances, be found useful."

The *Doncaster* was, of course, a sailing vessel, but in March 1803, we are told, the French organ builder, Charles Dallery (1754–1835), propelled a steam boat by means of a screw on the river Seine near Bercy. Of this experiment, however, we have little information. The credit for constructing the first practical screw-driven steam boat is usually accorded to Colonel John Stevens, the machinery of whose boat is preserved in the United States National Museum, Washington. In the succeeding quarter of a century many patents for screw propellers were taken out, but until the 'thirties there were few practical demonstrations. Among the more important inventors were Trevithick (1815), Millington (1816), Delisle (1823), Perkins (1824), Samuel Brown (1825), Ressel (1827), Cummerow (1828), Wilson, Woodcroft, Sauvage, Berthon, Blaxland, Smith, Ericsson and Lowe. Of these, Robert Wilson (1803–82), in 1827, drove a model boat with a screw and his success was acknowledged by the award of a medal by the Scottish Society of Arts, who brought the invention to the notice of the Admiralty. On the Continent, credit is given to the unfortunate French inventor, Pierre Louis Frédéric Sauvage (1786–1857), and to Josef Ressel (1793–1857). Sauvage's patent is dated 1832, but apparently, through lack of money, he was unable to put his ideas into practice. Before his death he became insane, but his merits had been recognised by Louis Philippe, and his statue now adorns Boulogne. Ressel's patent was taken out in 1827, and it was while he held a government post at Trieste that, in 1829, he fitted out the small vessel *Civitta* with an engine and propeller, which are described and illustrated in a centenary notice in *Schiffbau* for September 4, 1929. Ressel was a man of many parts, and he, too, is commemorated by monuments, one at Vienna and the other at his birthplace, Chrudin, in Czechoslovakia.

Without detracting in any way from the work of these

and other ingenious pioneers, it may, however, be said
that the adoption on a wide scale of the screw propeller
was due to the five inventors, Bennet Woodcroft (1803–
79), Francis Pettit Smith (1808–74), George Blaxland,
James Lowe (1798–1866), and John Ericsson (1803–
89), while in Smith's *Francis Smith* and *Archimedes*, and
in Ericsson's *Francis B. Ogden* and *Robert F. Stockton*, the
world became possessed of its first practical screw-

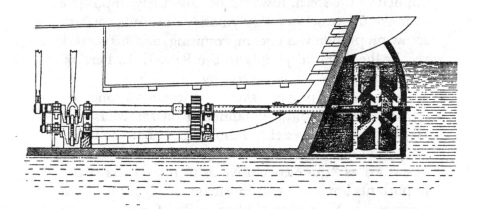

Fig. 3. Ericsson's screw propellers in *Robert F. Stockton*, 1838
From Woodcroft's *Origin and Progress of Steam Navigation*

driven vessels.  Smith's patent, the most important of
all, was taken out on May 31, 1836, and Ericsson's six
weeks later. Their screws were of totally different construction.  Ericsson's consisted of a short drum, like a
belt pulley, with blades around the circumference.  He
used two propellers, a right-handed and a left-handed
one, the outer, being fixed to the propeller shaft, revolving in a sleeve which carried the inner propeller.
The sleeve was driven by toothed gearing and the propellers revolved in opposite directions as they do in a
modern torpedo (see Fig. 3). Though Ericsson, with

the *Francis B. Ogden*, 45 ft. long, in 1837, towed the Admiralty barge, with a distinguished party of naval officers aboard, from Somerset House to Seaward's Works at Millwall, he received no encouragement, and a little later Sir William Symonds, the Surveyor of the Navy, expressed the opinion that "even if the propeller had the power of propelling a vessel, it would be found altogether useless in practice because the power being applied in the stern, it would be absolutely impossible to make the vessel steer". Symonds's attitude towards screw propulsion was uncompromising, and his remark recalls the Admiralty reply to the Rev. E. L. Berthon that "the screw was a pretty toy, which never would and never could propel a ship". Ericsson's second vessel, the *Robert F. Stockton*, launched on the Mersey in 1838, was an iron vessel 70 ft. in length. Both at Liverpool and London she carried out trials which *The Times* described as "quite conclusive as to the success of this important improvement in steam navigation", but Ericsson on November 1, 1839, sailed for America in the *Great Western*, and it was in the United States that he built his next ships with screws, among them being the *Princeton*, of the United States Navy. The only vessel in the British Navy fitted with Ericsson's screw was the *Amphion*.

Of the work of Smith many accounts have been given. For twenty years he devoted himself almost entirely to the subject of screw propulsion; it was mainly through his persistent efforts that the screw was introduced into both the naval and mercantile fleets of Great Britain, and his work won for him a world-wide reputation, the familiar name of "Screw" Smith, and finally a knighthood. How completely his contemporaries connected him with the adoption of screw propulsion was shown at a dinner in London on June 2, 1858, over which Robert Stephenson presided, when Smith was presented with an address, containing the names of practically all the most prominent shipbuilders and marine engineers, and a fine silver vase and salver, which he eventually bequeathed to the Patent Office Museum, now the Science Museum. These objects, which are preserved in the Marine Engineering Collection, have an additional interest for the student of history, for around the salver is a list of the names of all Her Majesty's vessels fitted with the screw up till 1858.

Smith was born on February 9, 1808, at 31 High Street, Hythe, a house which now bears a commemorative tablet, and died in South Kensington, February 12, 1874. He was buried in Brompton Cemetery. As a boy, he was fond of making working models, and he continued his hobby while a farmer at Romney Marsh and then at Hendon, and in 1835 he made a model screw-driven boat, which crossed the pond on his farm at the latter place. His patent of 1836 stated his invention to "consist of a sort of screw or worm made to revolve rapidly under water, in a recess or open space formed in that part of the after-part of the vessel, commonly called the dead rising, or dead wood of the run", while "the propeller might be of wood, sheet iron or other suitable material, and with a greater number of threads or worms, and set at various angles with the central line of the screw", and also that "it might be arranged singly in an open space, as there shown, or in duplicate with one on each side of the deadwood, or otherwise placed more forward or more aft, or more or less deep in the water". Three years later he limited his claims to a single propeller, as "I find that the dead wood or run of the vessel, is the only place in which the said propeller can be advantageously placed, and that a screw of one turn, or two half turns as a propeller will be sufficient for every purpose".

In the *Francis Smith*, a steam boat of six tons, built at Wapping in 1836, Smith had a wooden screw of one thread but two complete turns. During a trip on the Paddington Canal, a portion of the screw was broken off, but the speed of the boat increased rather than diminished. In 1837, this boat was fitted with a metal screw, and Smith made a trip from London to Ramsgate and Folkestone in rather rough weather. This success raised the hope of introducing the screw into naval vessels, and a syndicate was therefore formed called the Ship Propeller Company, to experiment with a larger vessel. Smith's patents were purchased, Henry Wimshurst (1804–84) was given an order for the vessel, while the Rennies were to supply the machinery.

It was under these circumstances that the *Propeller*, as the vessel was called at first, or the *Archimedes* as she was afterwards rechristened, came into existence (see Pl. IV). This historic vessel was 106 ft. long between perpendiculars, of 22½ ft. beam and of 240 tons. Her draught was from 9 ft. to 10 ft. The engines, of about 80 horse-power, had two cylinders, 37 in. diameter and 3 ft. stroke, and drove the propeller shaft through gearing, an engine speed of 26 revolutions per minute giving a propeller speed of 140 revolutions per minute. The boiler pressure was 6 lb. per square inch. Ready for trials in the spring of 1839, the *Archimedes* went from the Thames to Dover, Portsmouth and Plymouth, and then to Bristol, where Brunel was constructing the iron vessel *Great Britain*, and afterwards visited the principal home ports and made a voyage to Oporto. At Bristol, Brunel conducted trials with the *Archimedes*, and soon came to

PLATE IV

*ARCHIMEDES*
From contemporary engraving, 1839. Courtesy of Science Museum

H.M.S. *RATTLER* versus H.M.S. *ALECTO*, 1845
From contemporary engraving. Courtesy of Science Museum

the decision that the screw was preferable to the paddle-wheel, and without hesitation he abandoned paddle-wheel propulsion for the *Great Britain*, which ultimately became the first screw steamer to cross the Atlantic. When at Dover, the *Archimedes* had been pitted against the paddle-vessel H.M.S. *Widgeon*, the fastest boat on the Dover station, and the records of their respective performances, signed on May 2, 1840, by Captain E. Chappell, R.N., and Thomas Lloyd, was the first official report made to the Board of Admiralty on screw propulsion. Though the report stated that the power of the screw is equal if not superior to that of the ordinary paddle-wheel, and that the operation of the screw facilitated the steering, it called attention to the rapid wear and undesirable noise of the gear wheels, which, however, Smith proposed to get over by the substitution of spiral gearing.

The subsequent career of the *Archimedes*, after the conclusion of these trials, was a somewhat chequered one. She had already had a boiler explosion and a broken crankshaft when, in 1840, she went ashore off Beachy Head. Originally, she had been classed only 9 A1 at Lloyd's on account of having Baltic fir in her bottom. Her owners declining to carry out the repairs required by Lloyd's surveyors, her class was deleted from the *Register Book*. In 1845 she disappeared from the Register, only to reappear again in 1847, being classed 3 A1 after being overhauled. At Sunderland, her machinery was removed, and she became a sailing ship. Surveyed again in August 1852, she was given new sails and rigging, but a year or two later her class lapsed and her further history is unknown. Her original cost had been £10,500, and according to Woodcroft, her spirited proprietors lost some £50,000 in the venture. There appears to be no model of her in any nautical museum, but she nevertheless has a place in history beside such vessels as the *Clermont, Comet* and *Turbinia.*

The building of the *Archimedes* was speedily followed by the construction of other screw vessels, and no sooner was she ready for sea than Wimshurst built the *Novelty,* a somewhat larger vessel, which by her voyages from Liverpool to Constantinople with a cargo of 420 tons, became the parent of the screw-driven cargo boat. Other vessels built shortly afterwards were the *Great Northern,* constructed at Londonderry, the *Margaret* and *Senator,* built at Hull, and the *Princess Royal,* built on the Tyne. The *Princess Royal* was a pleasure boat for the South Coast traffic, and her first trip from Brighton to Arundel was made in June 1841.

Another early screw vessel was the *Mermaid,* of 164 tons, built at Blackwall and engined by Rennies. Fulfilling the condition that she should steam 12 miles per hour, she was taken over by the Admiralty, and, as H.M.S. *Dwarf,* became the first screw-propelled vessel in the Navy. The second screw vessel in the Navy was the *Bee,* a craft of 42 tons, built at Chatham in 1842, and fitted with a single-cylinder side-lever engine of 10 horse-power, driving both paddle wheels and a screw. Attached for many years to the Royal Naval College, Portsmouth, she was used for instructional purposes, and many curious experiments were made with her, some of which were even mentioned by Rennie in his Presidential Address to the Institution of Civil Engineers. On the opening of the Royal Naval College, Greenwich, in 1873, the machinery of the *Bee* was re-erected there, but at some time later it passed to the scrap-heap. In the beautiful little iron yacht H.M.S.

*Fairy*, with Penn's oscillating engines, the Navy also possessed an interesting early screw vessel.

Neither the *Dwarf* nor the *Bee* were fighting vessels, and the history of screw propulsion in warships begins with the construction of the *Princeton* for the United States Navy, of the *Napoleon* for the French Navy, and of the *Rattler* for our own Navy. Ericsson was responsible for the machinery of the *Princeton*, and Barnes for the machinery of the *Napoleon*, which was built at Havre by Normand, while the *Rattler's* engines were by Maudslays. Laid down at Sheerness in 1842 as the paddle vessel *Ardent*, the *Rattler* was altered for screw propulsion while on the stocks. No great haste was displayed in completing her, and she was turned over to Maudslays in the roughest possible state. Brunel, whom the Admiralty regarded as their adviser, left an amusing account of an interview he had with Admiral Cockburn about the *Rattler*, which illustrates the opposition Smith and his associates had to contend with. A sloop of 1078 tons displacement, the *Rattler* had a Maudslay "Siamese" engine of 437 indicated horsepower, driving the propeller shaft through gearing (see Fig. 4). Launched in the spring of 1843, she began her trials in October of that year, and these lasted till April 1845, when the famous tug of war took place. Her chief competitor was the paddle sloop *Alecto* of 800 tons and 200 nominal horse-power. They had raced each other in the Thames and off the East Coast until, on April 3, 1845, the two ships were lashed stern to stern. With the *Rattler's* engines stopped, the *Alecto* was allowed to go full speed ahead, but on the *Rattler* going ahead the *Alecto* was brought to a standstill, and finally was towed astern at the rate of two miles and a half an hour, "in spite of all her attempts to run away" (see

Pl. IV). Though lacking all the exactness of modern trials, and possessing little scientific value, this demonstration influenced popular opinion more than many pages of figures and of curves could have done, and

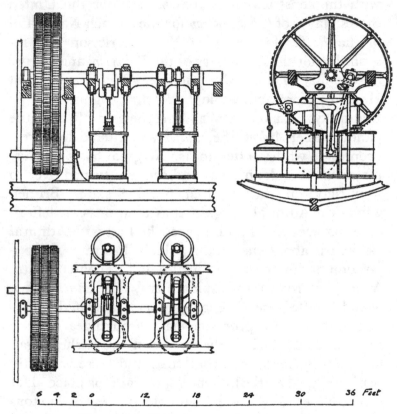

Fig. 4. "Siamese" engine for *Rattler*, 1843
Diameter of cylinder 40½ in. Stroke 4 ft.

from that trial dates the decision to adopt the screw for all classes of naval vessels.

As regards mercantile vessels, opinion on the relative merits of paddle and screw differed for some considerable time, but the superior advantages of the screw for naval vessels were acknowledged at once. The machinery could be placed below the water line and out of reach of shot; the decks could again be left clear for the armament; boarding enemy vessels again became an easy matter; the screw itself was not exposed to shot and the heeling of the ship when using steam alone, or steam and sail combined, did not affect the working of the screw as it did the paddle. Objections were raised on various minor points, but the trials of the *Rattler* definitely turned the scale, and though a few

### Table VII
### Screw Ships in Navy List January 1850

| | Displacement Tons | Guns | N.H.P. | I.H.P. | Engine makers |
|---|---|---|---|---|---|
| *Agamemnon* | 3750 | 91 | 600 | 1838 | Penn |
| *Ajax* | 3090 | 56 | 450 | 846 | Maudslay |
| *Amphion* | 2025 | 34 | 300 | 592 | Miller |
| *Arrogant* | 2615 | 46 | 360 | 774 | Penn |
| *Blenheim* | 2790 | 60 | 450 | 938 | Seaward |
| *Dauntless* | 2150 | 33 | 580 | 1347 | Napier |
| *Highflyer* | 1775 | 21 | 250 | 702 | Maudslay |
| *Hogue* | 3054 | 60 | 450 | 792 | Seaward |
| *James Watt* | 4950 | 91 | 600 | 1543 | James Watt & Co. |
| *Sans Pareil* | 3800 | 70 | 400 | 1471 | James Watt & Co. |

paddle frigates were afterwards constructed, screw propulsion was recognised as the most suitable plan for all classes of fighting vessels. By 1850 the Navy List included the names of about fifty screw vessels, particulars of a few of the larger of which are given in Table VII. Some of these vessels were old sailing ships converted to screw ships, the *Agamemnon* being the first line-of-battleship designed as a screw ship.

Effective as had been the trials of the *Archimedes* and *Rattler*, it was soon found that the new means of propulsion brought with it some perplexing problems. No easy way was found to the design of propellers, and tedious and expensive trials had to be made. Both *Archimedes* and *Rattler* had been fitted with various screws. Some of the screws tried in the *Rattler* are shown

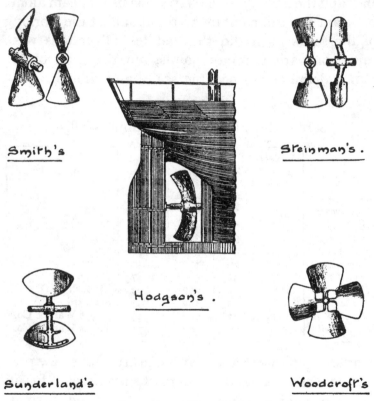

Smith's

Steinman's.

Hodgson's.

Sunderland's

Woodcroft's

Fig. 5. Screw propellers tried in *Rattler*

in Fig. 5. In 1845 the *Dwarf* was tried with 24 different screws, and a little later other trials were made in the gunboat *Minx*. Other screws than those of Smith were experimented with. Woodcroft had patented an increasing pitch propeller; Lowe had patented the use of segments of screws and Blaxland's screw was tried in a little craft called the *Jane*. Up till 1850, Smith worked under the Admiralty, and he and his contemporaries had been involved in long litigation. This, however, was brought to an end with an action in 1850, during the course of which Woodcroft declared: "The parties having patents for the screw propeller have united; they are Messrs Smith, Lowe, Ericsson, Blaxland, and myself. After fighting each other for many years, we have got tired of it and want to be amicable."

Of no less importance than the trials of propellers were the trials of ships with different forms of stern. To carry the heavy guns placed forward and aft, steam vessels had been built with full bows and sterns. In many vessels the bluff stern prevented the free flow of water to the propeller with resulting inefficiency. Scott Russell it was who once remarked that horse-power was not everything, and that shape largely determined speed. In the *Dwarf*, it was found that by altering the stern, the speed could be reduced from 9 knots to 4 knots. By altering the stern of the sloop *Rifleman* the speed was increased from 8 knots to $9\frac{1}{2}$ knots, while in the gunboat *Teazer*, by improving the stern it was found possible to obtain the necessary speed with far less horse-power.

When engine power was comparatively small, and the screw shaft was separate from the crankshaft, the thrust was usually taken by a plate opposed to the end of the shaft. The *Rattler's* thrust was of this description, and by a system of levers, a record of the variation in thrust was obtained. In the *Great Britain*, a gunmetal end to the shaft pressed against an iron plate 2 ft. in diameter over which played a copious stream of water. With direct-acting engines the propeller shaft was in direct line with the crankshaft, and the thrust was transmitted through the cranks. In the drawings of the arrangement in the *Ajax* there is a stout "pushing post" placed vertically, and fixed to keel and deck to take the thrust. A section of the *Ajax* is given in Fig. 6. Some-

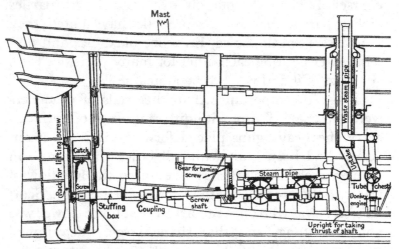

Fig. 6. Section of H.M.S. *Ajax*, 1846
From Bourne's *Treatise on the Screw Propeller*

times the thrust was taken on the stern post. Difficulties with such arrangements were bound to occur, and these led to the introduction of the multi-collar thrust placed between propeller and engine, which has been used exclusively until the introduction of the Michell thrust bearing.

Still more difficult of solution was the design of the underwater stern bearing, and the stern gland fittings were long a source of great anxiety. The copper sheathing and the iron shafts set up galvanic action; if shafts stood idle for any time there was rust; wooden ships worked in a seaway, and the shafts frequently were out of line. The early stern bearing was generally a long brass bush, and in vessels with copper sheathing the shaft had a brass sleeve. Bearing pressures were kept very low, but for all that the wear was abnormal, and in the sloop *Malacca* the brass of the stern tube wore away at the rate of 5 lb. a day. In the *Life of Admiral Sir W. R. Mends* is found a striking account of the voyage of the *Royal Albert* line-of-battleship, from the Dardanelles, when she had to be run ashore on the Island of Zea to prevent her sinking through the flow of water through the stern tube. She was kept with her bow ashore for four days, the engine slowly revolving to pump out the water, while the carpenters built a coffer-dam round the stern gland. This completed, the ship was towed off and then proceeded to Malta under sail.

The introduction of the lignum-vitae stern bearing was due to Penn, who was assisted by Smith, and this highly successful invention was the outcome of experimental research. On a shaft running in a tank of water Penn fitted bearings of various metals, alloys and wood, and he found that with iron working on lignum-vitae under water the bearing pressure could be carried up to 8000 lb. per square inch. The invention, which is described in the *Proceedings of the Institution of Mechanical Engineers* 1856 and 1858, came at a critical time; there was little delay in adopting it, and within two years 200 ships or more had exchanged their brass tubes for the new lignum-vitae bearings.

Two other important problems which occupied inventors were connected with the appliances for lifting the screw out of the water or for feathering the blades when a ship was sailing. Such fittings have long since become obsolete for vessels of any size, but in the 'fifties it was still generally thought that very long steam passages would be impossible, while for warships steam was long regarded as an auxiliary to sails and not a substitute for them. These views are reflected in the report of a paper "On the Application of the Screw Propeller to the Larger Class of Sailing Vessels", read before the Institution of Civil Engineers in 1855 by R. A. Robinson. The following year the Admiralty called the atten-

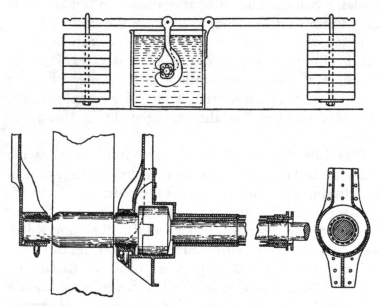

Fig. 7. Penn's experimental apparatus and his method
of fitting lignum vitae bearings

tion of commanding officers to the need for ensuring the efficiency of screw ships as sailing ships, and directed that steam power was not to be resorted to when the service on which the vessel was employed could be satisfactorily performed without it. The high fuel consumption had something to do with the issue of these orders, but there was an ever-present lack of confidence in the reliability of machinery, and the naval *Seamanship Manual*, in 1861, said "Engines and Machinery, liable to many accidents may fail at any moment, and there is no greater fallacy than to suppose that ships can be navigated on long voyages without masts and sails, or safely commanded by officers who have not a sound knowledge of seamanship", a statement a part of which is still another example of the failure of men of one generation to visualise what will be done by the men the next.

## CHAPTER VI

# NAVAL OFFICERS AND STEAM

T HOUGH fighting services are sometimes regarded as the strongholds of tradition and inertia,
they are probably neither more nor less so than
other long-established organisations. However this may
be, there are invariably individuals in them eager to
advance, to try out and adopt new devices and inventions. The success of such pioneers will depend much
on circumstances, actual warfare often furthering their
aims and hastening the application of their plans. In
our own day we have seen warfare revolutionised by
the internal-combustion engine, the submarine, the
mechanically driven vehicle and the aeroplane, and it
is unnecessary to emphasise the effect of the war of
1914–18 on the rate of progress. It was much the same
in the case of steam, the value of which was partially
demonstrated in various minor naval operations, and
more fully in the Russian War of 1854–56. More than
thirty years had then passed since steam was introduced
into the Navy, and in that time an increasing number
of naval officers had done their utmost to master the
problems of steam propulsion, and to study the probable changes it would effect in naval warfare. They
were the pioneers of their day, and no history of either
naval engineering or tactics would be complete without
a review of their work.

It must be admitted that when steam was first applied to the propulsion of boats, there were few circumstances calculated to assist the advocates of steam for
naval warfare. When Bell's *Comet* was launched on the

Clyde, we were still engaged in a long drawn-out war, and when a year or two later peace was declared, there could have appeared little necessity for experimenting with either new ships or new weapons. The war had been fought with ships and guns which had been evolved during the course of centuries, and it is in no way surprising that, when it was suggested that steam might prove of value in fighting ships, the older officers, who had seen the Navy emerge triumphant from its long struggle, should have looked askance at such a suggestion. Generally speaking, in naval circles there was almost an abhorrence of innovation, and one naval writer in 1826 said

the old school of seamen consisted, and still consists, of the most prejudiced beings in existence; nothing novel, in their opinion, was or is, either safe or available. They revered with a spice of idolatry, everything on the old plan, however tardy the process, or cumbrous the machinery, while they recoiled like a rusty cannonade at the very name of a novel invention, which affected either to lessen manual labour or promote despatch.

From the Admiralty, inventors certainly received little encouragement. Speaking of Lord Melville's second administration (1828–30), Sir John Barrow, the distinguished Secretary of the Admiralty, said:

The Admiralty, however, in this period of inactivity was beset with projectors of all descriptions. Steam vessels were fast increasing in numbers, some for public and others for private purposes; and all the gear appertaining to them, the engines themselves, the boilers, the mode of placing them, the paddle wheels, the paddle boxes, various kinds of propellers—all of them had a multitude of projectors, a class of persons who are never satisfied if each of their individual projects be not practically put to the test, however obvious it may be to the disinterested person, capable of giving a sound opinion, that the invention, as it is called, is bad in principle and worthless in design.

On another occasion, Barrow wrote of Lord Minto, the First Lord of the Admiralty in 1835, as possessing

a competent knowledge of the mechanical powers and of their various modes of application, which in these days of inventions, is no mean acquirement in a First Lord of the Admiralty, beset, as he is sure to be, by a host of speculative inventors, whom it is not easy to satisfy or get rid of, especially when they happen to be naval officers of high rank, who may fancy themselves capable of making improvements in naval construction, principally in steamers, of which they can have but a very imperfect knowledge.

If there was one officer in particular whom Barrow had in mind when he wrote, it must have been Captain (afterwards Admiral Sir) Charles Napier (1786–1860) who in 1828 had asked in vain for the command of the *Lightning*, and in 1832 wrote to Sir James Graham, then First Lord, stating,

It does appear to me, that neither the engineers who make the engines, nor the builders, who construct the vessels, have the least idea of what is necessary to constitute a steam man of war...if you will permit me to fit one of them out I should have no objection in showing the builders and engineers how the vessel and engines should be constructed and placed so as to render them secure.

When given the command of the sailing vessel *Galatea*, instead of the *Lightning*, Napier immediately fitted her with paddle wheels driven by winches on the upper deck worked by the crew, and with her wheels in action took her out of Portsmouth harbour. As early as 1822, he had navigated the little iron steam vessel *Aaron Manby* from London to Paris, and he claimed that he had given more attention to steam navigation than any other officer. Very outspoken in his criticisms, he wrote *The Navy, its Past and Present State*, and in this and his *Memoirs and Correspondence* are to be found many interesting facts regarding our early steam Navy.

Napier, however, was not the first naval officer to write on steam. That distinction belongs to his contemporary, Captain (afterwards Rear-Admiral Sir) John Ross (1777–1856) who, like Napier, had seen service afloat during the Napoleonic wars. A few years after peace had been declared, Ross had made his first voyage to the polar regions. His voyage had not been a very successful one, but later on he was enabled to fit out another vessel for the same purpose, and this time he chose a steamer, the *Victory*, into which Braithwaite and Ericsson fitted high pressure boilers, a surface condenser and forced draught apparatus, when such things were novelties. If space permitted, it would be diverting to recall the extraordinary story Ross had to tell in his *Narrative of a Second Voyage in Search of the North-West Passage; during the years 1829–1833*. It is only recently that details and sketches of the machinery have been found, and these show that the *Victory's* boiler was of the same type as that Braithwaite and Ericsson made for the locomotive Novelty, which competed with the Rocket in the Rainhill trials on the Liverpool and Manchester Railway in October, 1829. For locomotive work this type of boiler proved fairly successful, but the machinery of the *Victory* was a complete failure, and at the very time of the Rainhill trials the crew of the *Victory* were removing the engines and boilers from the ship and placing them on the ice in latitude 70° N., where they were left.

Ross had undoubtedly paid much attention to steam, and his *Treatise on Navigation by Steam; and an Essay towards a System of Naval Tactics peculiar to Steam Navigation*, published in 1828, the year before he sailed north, was the forerunner of many somewhat similar works by naval officers. Besides chapters on the steam engine and steam ships, Ross included others on tactics, and the importance of establishing regulations for steam navigation. His views on the type of machinery which should be fitted in steamers making long voyages are given, and he says, "it is manifest that the machinery which is the least complicated, and least liable to derangement, which affords the most power in proportion to the space it occupies, which possesses the least weight, which is most speedily brought into action, and which consumes the least fuel is that which ought to be adopted for this service". Far in advance of his time, Ross was an ardent advocate of both water-tube boilers and high-pressure engines, the adoption of which, however, he did not live to see.

Of other works by naval officers to which we are indebted for information on early marine engineering practice, and the application of steam to naval warfare, reference has already been made to the *Elementary Treatise on Steam* by Commander R. Otway, and the *Memoirs of Her Majesty's Steam Ship the Medea* by Lieut. T. Baldock. These books were followed by the *Nautical Steam Engine Explained and its Power and Capabilities Described for the Use of the Officers of the Navy*, published by Commander R. S. Robinson in 1839; a *Steam Manual for the British Navy*, by W. J. Williams, Captain, R.N., 1843; *The Economy of the Marine Steam Engine*, by W.

Gordon, Lieut., R.N., 1845; *The Screw Fleet of the Navy*, by E. P. Halsted, Captain, R.N., 1850, and *A Treatise on Economy of Fuel*, by Alfred P. Ryder, Captain, R.N., 1852. Captain Williams's book was the outcome of some lectures given in the Royal Naval College, Portsmouth, where many officers were afterwards taught the rudiments of steam by Chief Engineer Thomas Brown, the joint author with Professor T. J. Main of *The Marine Steam Engine, Designed Chiefly for the Use of the Officers of Her Majesty's Navy, 1847*, and of *The Indicator and Dynamometer*. Naval officers also took part in discussions at the Institution of Civil Engineers, and in 1857 Commander L. S. Heath contributed a paper to the Institution *On the Nominal Horse-Power of Steam Engines*.

The mere mention of these works is sufficient to show that the interest taken in steam propulsion was widespread in the Navy, and probably more so than is generally realised. Further evidence on this point is, moreover, to be found elsewhere. In the miscellaneous writings and the biographies of eminent officers, the reader will often learn something of the steps taken by officers to prepare themselves for the coming change and of the difficulties they struggled with. Captain Basil Hall, writing in the *United Service Magazine* in 1839, impressed upon the rising generation of officers the duty of turning their attention to steam. He told them to take off their jackets, put on paper caps, be regardless of soiling their fingers, and to submit "to the companionship of thoroughbred workmen in Maudslay's works in London or those of Napier in Glasgow". Napier's was a favourite place for learning something of the mysteries of the workshop, and Napier's hospitality was proverbial. In 1843 we find Captain

Erasmus Ommanney, of the *Vesuvius*, writing to Napier from the Mediterranean as one of "your old apprentices", and, as late as 1851, Admiral Sir Thomas Cochrane wrote to Napier on behalf of Lord John Hay, then a commander, who wished "to attend and take advantage of the scientific instruction he can receive". Seaward's was another works visited by officers. Some officers, too, gained experience by taking voyages in mercantile steamers. Admirals Sir Astley Cooper Key, Sir William Mends, Sir William Hutcheon Hall, and Sir Edward Fanshawe, in their early days, all studied steam when the subject was regarded with a certain degree of bewilderment, and even suspicion.

Commander Robinson, in his book, helps us to understand some of their perplexities. He remarks somewhat pathetically,

we go into the engine room, we look at the outside of an engine, various rods of highly polished iron are moving about, a beam is observed vibrating up and down, all is clean and bright and well arranged, but the working parts of the engine, the moving power is entirely shut out from our sight, and after staying a few minutes and, perhaps, asking a question or two, which from the very depths of ignorance it betrays, it is scarcely possible the engineer can or will answer, we walk up again, with no additions to our knowledge, and rather convinced that the whole subject is incomprehensible.

Robinson himself, as his book shows, mastered the subject, and he was in no uncertainty as to the value of a knowledge of navigation by steam. After explaining the object of his book, he goes on to say

It is evident that whatever of dash, whatever of enterprise, whatever of combined prudence and skill, is to be performed in a future war will be performed through the agency of steam. The high road to distinction and fame will be found on the paddle box of a steamer, but to gain this fame, to achieve this distinction, it is indispensable that officers should add to a thorough knowledge of seamanship and gunnery, to nerve, to enterprise, to prudent daring, a knowledge of the steam engine, an acquaintance with the power which is to be their right arm and their strong staff.

Robinson's views were not those generally held among older officers for, when Cooper Key, in 1844, exchanged from the sailing vessel *Curaçoa* into the steam frigate *Gorgon*, current opinion in the messes, as Colomb tells us in his biography of Key, thoroughly despised the growing propulsive power. "Steamers' midshipmen" and "steamers' lieutenants" were terms of opprobrium in the rising generation, and even the captain and commanders who commanded steamers were regarded as a lower grade of officer, lost to the traditions of the service, and out of the running in the matter of distinctions.

As time went on, facilities for officers desirous of studying steam were afforded by the Admiralty at Portsmouth and Woolwich, and regulations were laid down for their guidance. An order in 1841 had already directed that, in addition to the practice of naval gunnery, marine officers studying at the Royal Naval College "shall henceforth acquire a thorough knowledge of steam machinery"; and in a memorandum of December 8, 1849, the Admiralty directed that no officer should henceforth be registered as a "steam officer" who had not passed an examination at the Royal Naval College, whether he had acquired his knowledge in one of H.M. Dockyards or at the factories of private individuals or had studied at the Royal Naval College or not. The examinations included geometry, dynamics, heat and combustion, and the principles of engines and boilers, and they appear to have been in use until, at least, the time of the Russian War.

It was but natural that the naval officers whose works have been mentioned should be as much concerned with the effect of steam vessels on tactics as with the development of the engines and ships themselves. Ross in his work of 1828, when writing of steam navigation, declared that "the adoption of this mode of motion, and these new inventions will produce an entire revolution in the present system of attack and defence, and that an entire new method of tactics must be a necessary consequence". A considerable part of his book is taken up with discussions of the handling of steam vessels under varying conditions of weather, of the regulations it would be necessary to introduce, and of steamers as auxiliaries to the sailing vessels, as separate fighting forces, as convoys for merchant vessels and for coast defence.

Napier, too, urged upon the Admiralty not only the necessity of providing steam vessels—the cavalry of the Navy—as auxiliaries to the Fleet, but also of the desirability of providing the larger vessels with their own means of propulsion, even if these were hand-worked. If we were to continue masters of the sea, our steam-boats ought to be off every port of the enemy where they have steam vessels ready to take advantage of calms, but he said "as each ship of war cannot be accompanied by a steamboat to take care of her in a calm, I cannot see why 600 or 700 men are to remain to be shot at, either ahead or astern, when they have the means, at a very trifling expense, of moving between 3 or 4 knots and thus keeping their broadsides to the steamers".

The idea of each line-of-battleship being accompanied by a steam vessel appears to have taken root in the 'forties, and is found expressed in the official correspondence given in Admiral Phillimore's *Life of Sir William Parker*, the Commander-in-Chief of the Mediterranean, and in other books. The letters quoted below were written just when the screw was coming into use, and the relative merits of paddle and screw and the use of steam vessels in a fleet were being discussed.

In one letter to Sir William Parker, written in 1846, the Earl of Ellenborough, who previously had been First Lord, said

I am glad to find the paddle steamers with their floats off can keep company with the fleet, but it is very necessary that means should be devised of enabling them to re-ship their floats rapidly so that they may at once resume their characters as steamers. It is as steamers only that they would be of use in action, and actions come on sometimes so suddenly, that unless they can rapidly resume their proper character, they would be of very little, or no, use. I agree with you in thinking that the paddle steamers will give place to the screw steamers for all service with fleets.

Writing in March 1847, Lord Auckland, who had succeeded the Earl of Ellenborough, wrote, "You know how favourite a notion it is with me that every strong ship should have her attendant steamer, and I think that your squadron may consist of

| | H.P. | | H.P. |
|---|---|---|---|
| *Hibernia—Terrible* | 800 | *Canopus—Dragon* | 560 |
| *Trafalgar—Retribution* | 800 | *Superb—Centaur* | 540 |
| *Rodney—Sidon* | 560 | *America—Gladiator* | 430 |
| *Albion—Avenger* | 650 | *Thetis—Scourge* | 430 |

I attach infinite importance", he added, "to the combined movements of steamers and sailing vessels, not only for mutual support, but also in the economy of means by the management of fuel".

About the same time the Channel Squadron under Napier, then a Rear-Admiral, actually consisted of the five line-of-battleships, *Howe, Caledonia, St Vincent, Queen* and *Vengeance*, and the five paddle steamers, *Odin, Dragon, Vixen, Stromboli* and *Avenger*.

Admiral Parker's letters home showed he agreed with Lord Auckland, for in April 1848 he wrote,

With regard to steamers, I think there should be at least one for every line-of-battleship, besides what will be necessary for detached services...but these steamers with the tubular boilers must be well looked to, for without alteration, owing to flame emitted from the funnels, they are not safe to tow alongside; and if an attack is made on a force at anchor or in heavy ships in a calm, where towing is necessary they ought to be lashed alongside or they may be sunk, or disabled in approaching the enemy.

In October of 1849, Parker told the Admiralty of the passage of the Fleet to Besika Bay. "The wind was adverse from the time we rounded Cape St Angelo, but the weather being moderate the *Dragon* towed the *Howe* and *Caledonia* together through the Douro Channel, effecting a speed of five knots an hour with the two three-deckers, while the wind was light. I was very glad to obtain this result of her power." The utility of steam vessels for towing was shown again and again. Ships were towed into action in the Black Sea campaign, and years afterwards the flagship—a line-of-battleship—of the North American Station was accompanied by a paddle ship, which towed the battleship during calms and light head winds, but which was herself towed during fresh and fair winds in order to save her coal.

In the correspondence with Admiral Parker, Lord Auckland frequently referred to the new screw ships. In July 1846 he wrote to Parker,

I trust that you will have steamers enough to throw further light on the comparative merits of screw and paddle. I see a very growing opinion in favour of the screw and clearly for many reasons, it has great advantages, but I am not yet satisfied of its superiority under the direct action of steam power, or of its being less subject to the want of frequent repair.

A month later Parker replied,

We have also tried the power of the *Rattler* and *Polyphemus* in which the former prevailed when stern to stern, and she had a decided advantage in a steam run of eighteen miles in a calm; but when lashed bow to bow on a stern-board the *Polyphemus'* paddles had the best of it. The *Rattler* performs to admiration and keeps company like a frigate under canvas only. *Polyphemus* also does well with her floats unshipped, and with additions to her rig might be still improved.

A few days later Lord Auckland replied,

What you say of the performances of the *Rattler* is very satisfactory and important. Immense sums have been expended on the construction of steamers and machinery (and necessarily so) long before the best forms and sizes of vessels and the best description of propulsion could be satisfactorily determined, but I think we are beginning to learn something definite by experience, and I am led to believe that very shortly the screw will in a great measure supersede the paddle.

In November he wrote,

I find some high authorities, and particularly with Sir W. Symonds, a strong disposition to underrate the screw steamers, and I am slow in adopting their objections. I have no doubt that the paddle is the more powerful engine and the least liable to get out of order. But the screw is safest from shot and from its facilities for disconnecting, if the vessels are good for sailing may be in a great degree preferable for cruising. Let me hear from you on all this.

Ten years later the highest naval authorities were discussing, not the relative merits of the paddle and screw, but the manœuvring of whole fleets under steam for it was then realised that, as General Sir Howard Douglas remarked in his work *On Naval Warfare with Steam*, 1858, "Steam propulsion entirely annuls the limitations and disabilities imposed by the wind on the evolution of fleets, and opens the whole surface of the ocean as a battle-field for the contests of steam fleets".

CHAPTER VII

# IRON SHIPS FOR MERCANTILE PURPOSES

WHILE the general history of shipbuilding does not come within the scope of this book, the early development of iron shipbuilding was so largely the work of mechanical engineers, and the coming of the iron ship had so great an influence on the progress of naval and marine engineering, that a brief review of the early history of the iron steam ship and the iron warship cannot be out of place. Just as it was the work of Newcomen, Watt and others in connection with the steam engine, which led to the rise of mechanical engineering, so it was the epoch-making discovery by Henry Cort in 1783, of the processes of producing wrought iron by puddling and of rolling it into plates and bars which provided the new race of engineers with the material for their boilers, their ships, their railways, and their bridges. It was the art of boiler making which led to the art of iron shipbuilding, and thus it came about that nearly all the early iron ships were constructed by mechanical engineers and not by the established shipbuilders, and that nearly all of them were steam ships and not sailing ships.

In its general features the story of the iron ship presents much the same characteristics as that of the steam ship or the railway. There were similar pioneering experiments and similar practical demonstrations, and the same scepticism had to be overcome. Iron shipbuilding owed nothing to official encouragement, and its rise was entirely due to private enterprise. The objections, real or imaginary, were quite as numerous as those raised against railways. If iron vessels did not sink under their own weight, the engines, it was said, would cause them to sag, while the reciprocating motion of the engines would cause fracture of the plates. The differences in temperature of the two sides of a ship caused by the sun shining on one side, would cause serious stresses to be set up, while nothing would prevent rapid corrosion of the hull. Because it was found that shot fired through iron plates splintered badly, iron was declared totally unsuitable for men of war, and inasmuch as certain mercantile vessels were to be available for fighting, a ban was placed upon iron as the constructional material for vessels carrying mails. In official circles, too, there was a fear that the introduction of iron warships would be greatly to the disadvantage of the Royal Dockyards, from which came the great wooden line-of-battleships.

The most difficult problems which arose in connection with iron ships were due to the serious disturbance of the compass and the rapid fouling of the hulls by marine growths. By the 'forties practically all steamboats for river work in this country were built of iron and in these the problems mentioned were not of serious consequence. It was otherwise with iron ships employed on cross-channel and ocean voyages, for the disturbance of the compass endangered the navigation of the ships, and the fouling seriously reduced their speed. It was, indeed, the fouling which for many years hindered the adoption of iron for ships making long voyages. But with all these disabilities, the iron ships proved to be lighter and faster than wooden ships; they afforded more accommodation; cargoes could be more easily stowed and goods were found to arrive at their destinations in better condition. Experience also showed that iron ships were not so easily damaged; that repairs could be more readily effected in the yards, and with the introduction of bulkheads and double bottoms, iron ships became far safer than wooden ships. Then, too, iron was becoming more and more abundant, while timber was becoming increasingly scarce, so that whereas a wooden ship cost about £16. 10s. a ton, an iron ship could be built for £13. 10s. a ton.

Official recognition of iron ships may be said to date from the classification by Lloyd's Register of the steamer *Sirius*, 180 tons, built at Millwall by Fairbairn in 1837. The second iron ship classified was the sailing vessel *Ironsides*, 270 tons, built at Liverpool in 1838. In 1843 the Society determined to collect evidence regarding the building of iron vessels, and in 1844 granted the classification Iron ship A1 to such ships as were built under the inspection of the Society's surveyors. No rules for the construction of iron ships were issued till 1855. The germ of the extensive system of Government supervision of steam ships now in operation was an Act of Parliament of 1846, passed for "regulating the construction of sea-going steam vessels and for preventing the occurrence of accidents in steam navigation", and this Act directed that all iron steam vessels of 100 tons burden and upwards should be divided by watertight partitions into three compartments by transverse bulkheads at the fore part of the engine-room and at the after part.

Like the first cast-iron bridge, the first wrought-iron boat was built in Shropshire. Abraham Darby, the second, in 1779 had erected the famous Coalbrookdale Bridge over the Severn—a fine piece of work still standing—and it was his contemporary, John Wilkinson, the

ironmaster, who in 1787, at Willey, lauched the iron
canal barge generally spoken of as the first iron boat.
This craft was 70 ft. long and could carry 32 tons. It
had wooden stem and stern-posts, but its hull was of
$\frac{5}{16}$-in. wrought-iron plates. Other similar craft for
canal work were built shortly afterwards. Twenty years
later, Trevithick and Dickinson, in a patent of 1809,
described iron ships of war, iron East Indiamen, and
large iron-decked vessels. They even proposed iron
masts and yards. In 1815 a small iron boat, built at
Tipton, was sailing on the Mersey; and in 1818 the
*Vulcan*, an iron passenger boat built of sheets of iron
riveted to frames of flat bar iron, was set to work on the
Forth and Clyde Canal. The first iron steam vessel was
the *Aaron Manby*, which Napier navigated from the
Thames to the Seine in 1822. Built at the Horseley
Ironworks at Tipton, in the Black Country, the *Aaron
Manby* was sent in sections, by canal, to the Surrey
Docks, Rotherhithe, and there put together and fitted
with an oscillating steam engine and with feathering
paddle wheels. Declared to be "the most complete
specimen of workmanship in the iron way that has ever
been witnessed", she was named after the ironmaster
who successfully laid the foundation of the engineering
works at Charenton and Creusot, in France, and was
built in connection with his scheme for a steam boat
service on the Seine.

Among the principal pioneers of iron steam boat
building in this country were Grantham, Laird, Fairbairn, David Napier, Thomas Wingate, Ditchburn,
Mare, and the partners David Tod and John Macgregor, who were the first on the Clyde to make a business of iron shipbuilding. Like the *Aaron Manby*, the
second iron steam boat also came from the Horseley

Ironworks, but was put together at Liverpool. She was constructed in 1825 for the elder John Grantham and had a 10-horse-power engine. As the *Marquis of Welles-ley* she ran for many years on the Shannon. Next to her came David Napier's *Aglaia*, 49 tons, with an iron bottom but wooden sides above the water-line, built for work on Loch Eck, in Scotland. Laird began building boats of iron in 1829, and in 1831 constructed the iron steamer *Alburkak*, 70 tons and 15 horse-power, which accompanied Macgregor Laird to the Niger. About the same time John Neilson built the *Fairy Queen*, 40 tons, the first iron steam boat to ply on the Clyde, and at Manchester Fairbairn built the *Lord Dundas* and the *Manchester*. In 1832 Maudslay, Sons and Field constructed four iron paddle steamers, the *Lord W. Bentinck*, *Thames*, *Megna*, and *Jumna*, for work on the Ganges. These were shipped to India in pieces. They were 120 ft. long, and of 275 tons, with engines of 30 nominal horse-power, and they were the first iron steamers built in the London district.

In the 'thirties iron shipbuilding became firmly established on the Mersey, Clyde, and Thames. Laird's *Alburkak* was followed by the *John Randolph*, 250 tons, sent to America in pieces; the *Lady Lansdowne*, 130 ft. long, put together in Ireland; and the *Garryowen*, 263 tons, 80 horse-power, also built for Ireland. At the request of Charles Wye Williams, Laird fitted this vessel with bulkheads. Twice the *Garryowen* was stranded on the rocks, but in 1866, after running for years on the Shannon, she was still worth buying for work on the coast of Africa. It was in this vessel that, in 1835, Captain Johnson, by direction of the Admiralty, made experiments with compasses, but his investigations were not so important as those made in 1838 by Airy, the

7-2

Astronomer Royal, aboard the *Rainbow* in Deptford basin. The *Rainbow*, of 580 tons, was also built by Laird, and belonged to the General Steam Navigation Company.

On the Thames, iron shipbuilding received a great impetus by the establishment of the yards at Millwall belonging to Fairbairn and David Napier. Fairbairn launched about 100 iron vessels in the Thames, but, it is said, lost £100,000 in the undertaking, which had to be made good by the profits of his works at Manchester. His first iron vessels, as well as those of Napier and of Ditchburn, of Blackwall, were launched in the later 'thirties.

No constructors did more for iron shipbuilding in its early days than Tod and Macgregor of Glasgow. Both partners had previously worked for David Napier, but, setting up as engine builders for themselves, they added shipbuilding to their activities, and in 1835 launched the *Vale of Leven*, 121 tons, and the *Royal Tar*, 141 tons, in 1837, and these were followed in 1839 by the *Royal Sovereign*, 177 ft. long, 446 tons and 220 horse-power, and the *Royal George*, of about the same size, for the traffic between Liverpool and Glasgow. In 1842 they also built for the same route, the *Princess Royal*, 194 ft. long, the success of which went far to remove the prejudice against iron steamers. This was the vessel which was selected by a Committee of the House of Commons when inquiring into the matter of conveying mails to Ireland, and she, with her sister vessels and Laird's *Rainbow*, may be taken as the precursors of the ocean-going iron steamer.

The gradual supersession of wooden steam boats by iron steam boats for short-distance traffic took place during the two decades 1830 to 1850; the supersession of ocean-going wooden mail steamers by iron steam ships during the decades 1840 to 1860. In the 'forties and 'fifties a few iron steam warships were constructed, but iron as a material for fighting ships made little headway till after the Russian War. River craft, though numerous, were, of course, comparatively small, and thus of new shipping launched in this country in 1850, only one-tenth was of iron; in 1860, when iron sea-going ships were being built, this proportion had risen to one-third.

To illustrate the progress of the iron ship in both mercantile and naval fleets, reference is made here to some of the most notable vessels, built, respectively, for transatlantic work and for traffic to the Far East, while in the next chapter some notes are given on iron ships for fighting purposes or as auxiliaries. Each of the vessels selected for notice either embodied some new and important principle in her construction, represented some notable advance, or possessed some historical associations of permanent interest.

The first iron steam ship built for and employed on the Atlantic, was Brunel's *Great Britain*. No sooner had the *Great Western* achieved her initial successes than plans were drawn up for a still larger vessel, the very size of which made the adoption of iron almost a necessity. Originally called the *Mammoth*, a name justified at that time by her dimensions, like the *Great Western*, she was built at Bristol, by Patterson. The *Great Western* had been launched on July 19, 1837. Exactly two years later, on July 19, 1839, the keel of the *Great Britain* was laid down in a dry dock, and she was floated out July 19, 1843. Contemporary iron vessels were of about 700 tons to 1000 tons. The *Great Britain*, at a draught of 18 ft., had a displacement of 3618 tons, being 322 ft. over

all, 289 ft. between perpendiculars, 50½ ft. beam, and
32½ ft. deep. Her keel was of flat iron plates, ⅞ in. thick,
20 in. wide, welded into lengths of 50 ft. The skin
plating varied in thickness from ⅜ in. to $\frac{11}{16}$ in., the plates,
6 ft. by 3 ft., apparently being the largest then obtainable. Her frames were from 18 in. to 24 in. apart, and
five transverse bulkheads divided the ship into six compartments. Her machinery was as remarkable as the
ship herself, for she was the first large vessel to be driven
by a screw, and many new problems had to be solved.
Originally she had been intended to be driven by paddle
wheels, and it was Brunel's experiments in the *Archimedes* that led to the alteration in the plans. A model of
her engines in the Science Museum shows four direct-
acting cylinders, each 88 in. diameter and of 6-ft.
stroke, placed at an angle of 33° with the vertical,
driving an overhanging crankshaft which carried a
drum 18¼ ft. diameter and 38 in. wide, which by means
of four flat pitch chains, drove a smaller drum about
6 ft. diameter, fixed on the propeller shaft. Steam was
supplied by flue boilers working at 5-lb. pressure, and
at 18 revolutions per minute the engines developed
about 2000 horse-power, giving the vessel a speed of 12
knots. Having visited London, where she was inspected
by Queen Victoria, the *Great Britain* left Liverpool on
her maiden voyage to New York on July 26, 1845, and
crossed and recrossed several times during the next
twelve months. On September 22, 1846, when on an
outward voyage, she ran ashore in Dundrum Bay, Co.
Down, Ireland, and by the way she withstood the
storms of the following winter, gave convincing proof
of her great strength. Through the ingenuity of Brunel
and the exertions of Bremner, the salvage expert of the
time, on August 27, 1847, she was refloated, but was

never employed on the Atlantic again. Without the funds to repair her, the Great Western Steamship Company, to whom she belonged, sold her to Gibbs, Bright and Company, of Liverpool, and after being fitted with new machinery made by Penn in 1852, she began a long and useful career as an auxiliary screw ship on the England to Australia run. In 1882 she again changed hands, was converted into a sailing vessel, and finally was used as a hulk at the Falkland Islands.

While the *Great Britain* was preparing for her maiden voyage, another iron steamer, the *Sarah Sands*, was being built at Liverpool by Hodgson and Company, to the designs of John Grantham. She was intended to demonstrate the use of auxiliary steam power for general trading purposes on the Atlantic. Though in the end the plan of using steam solely as an auxiliary proved unsuccessful commercially, it was tried over and over again, on account of the high coal consumption of marine engines of the time. The *Sarah Sands* was 220 ft. long and of 1300 tons, with engines of 200 horse-power, and in 1847 she went from Liverpool to New York in 20 days. Some of her later voyages, however, were much shorter. After running some time on the Atlantic, she was taken over as a transport, and, in August 1857, left Portsmouth for Calcutta, with 300 troops aboard. On November 11, when 400 miles from Mauritius, she was found to be on fire aft. The ship, however, was efficiently divided by bulkheads, and by throwing streams of water onto the after bulkhead, the fire was confined within the after compartment, and after 16 hours' exertions, was put out. A storm which then arose flooded the engine room, but in spite of all, the vessel was navigated to Mauritius, under sail, without loss of life, her escape being as fine a testimony to the seamanship displayed as to the soundness of the construction of the ship.

Regular transatlantic passenger traffic by iron steamship was inaugurated by the *City of Glasgow*. This vessel was built by Tod and Macgregor, with the object of connecting Glasgow and New York. With accommodation for 600 passengers, her length was 237 ft., her beam 34 ft., and her tonnage 1670. Her screw engines were of 350 nominal horse-power. Like the *Great Britain*, she had five transverse bulkheads. Having made one or two voyages in 1850, she was purchased by William Inman, and became the pioneer ship of one of the most famous Atlantic steamship companies. A somewhat larger vessel, the *City of Manchester*, also built by Tod and Macgregor, was the second ship of the line, which for many years contended for the blue ribbon of the Atlantic. The career of the *City of Glasgow* was unfortunately a short one, for after running successfully until 1854, on March 1 of that year she sailed for America with 480 persons aboard, and, like the *President*, was never heard of again.

The work of Inman's ships and the growing competition of other companies were not without their influence, and Government opposition to the use of iron for mail steamers gradually disappeared. All the Cunard mail ships had necessarily been of wood, to comply with the requirements of the Government, but in 1854 Robert Napier built the *Persia*, the launch of which marked a new advance in Atlantic liners. This vessel was 390 ft. long, 45 ft. wide, 3600 gross tonnage, with engines of 4000 indicated horse-power. Kirkaldy's fine sectional drawings of the ship had the unique distinction of being exhibited at the Royal Academy. After the launch of the *Persia*, American shipbuilders continued to construct ocean-going steamers of wood for a time, but no such ships were subsequently built in Great Britain.

While these changes were taking place in the character of Atlantic liners, similar changes were being made in vessels built for the traffic to the Far East, and it was this traffic which led to the construction of the most famous of all nineteenth-century steam vessels, the *Great Eastern*. Until the opening of the Suez Canal in 1869, steam communication with India and China involved the overland journey from Alexandria to Suez. From Suez to Bombay, passengers proceeded in steamers owned by the East India Company, but passengers proceeding to Ceylon, Calcutta and Hong Kong were carried in vessels of the Peninsular and Oriental Steam Navigation Company. This company was one of the first to adopt both iron ships and screw ships, one of the earliest of them being the *Pottinger*, built by Fairbairn at Millwall. Other iron P. and O. steamers built in the 'forties were the *Ripon* and *Euxine* and the *Sultan*, a screw ship constructed by Tod and Macgregor in 1847, 231 ft. long, 32 ft. wide, and of 1124 gross tonnage. Her engines were by Caird, of Greenock. These vessels were followed in the 'fifties by the *Pera*, *Colombo*, and then the *Himalaya*, which bore comparison with the *Persia*. Built by Mare at Blackwall, in 1853, the *Himalaya* was 340 ft. long, 46 ft. wide, and had a displacement of 4690 tons. Her trunk engines by Penn, of 2500 horse-power, drove a single-screw 18 ft. in diameter. In the early part of the Russian War, she made a remarkable passage with troops from Gibraltar to Malta, averaging $13\frac{1}{2}$ knots. Afterwards, as H.M.S. *Himalaya*, she had a long and useful career as a naval transport, and then was relegated to the harbour service as a coal hulk. In passing, it is of interest to recall that on one occasion the *Himalaya* required repairs to her stern tube. No dry dock at Malta being large enough to accommodate her, she was floated into the largest available, and then by transporting weights to her forward end and by means of ropes, passed under the ship and then to capstans manned by soldiers from the garrison, her stern was raised out of the water sufficiently high for the repairs to be carried out.

The voyages to Alexandria and from Suez to the East, once the coal supply was arranged for, presented no great difficulties. It was otherwise with steam communication with the Cape and Australia, and in their attempts to establish regular steamship passages to Australia, several companies sustained heavy losses. The Government first asked for tenders for conveying mails to the Cape by steamship in 1850, and the contract was given to the General Screw Steam Shipping Company. The vessels employed were the iron screw ships *Bosphorus*, *Hellespont* and *Propontis*. These vessels, built at Blackwall by Mare and engined by Maudslays, were only 175 ft. long and 560 tons, but the sailing of the *Bosphorus* from Plymouth, on December 16, 1850, was regarded in much the same way as the departure of the *Britannia* from Liverpool had been just ten years before. Another iron steamer of the same company, the *Harbinger*, is of some interest inasmuch as she had commenced her existence as the *Recruit*, the only iron sailing vessel built for the British Navy. To adapt her for the screw, the *Recruit* was cut through amidships and drawn apart 76 ft. for the fitting of a new section, then both the forward and after parts were found unsuitable and were discarded, and all that finally remained of the *Recruit* in the reconstructed *Harbinger* was

70 ft. of her upper deck. But neither the *Bosphorus* nor her sister ships solved the problem of steam communication with the Cape; the venture failed, the vessels were taken up as transports, and some of them were afterwards converted to sailing ships.

To steam to Australia was a still more difficult task than to steam to the Cape, but the direct passage was accomplished in 1852 by the *Australian*, built of iron by Denny at Dumbarton and owned by the Cunard Company. She was 236 ft. long, 34 ft. broad, and her geared beam screw engines indicated 576 horse-power. Despatched from Plymouth, June 5, 1852, she did not arrive home again till January 1853, her round voyage having taken 221 days. Machinery troubles, frequent coalings and other matters had delayed her considerably, and she was not placed on the run again. The *Great Britain* was one of the vessels which followed her, and though there was some interruption with the Australian and other services during the Russian War, steam communication was maintained more or less regularly.

These observations on the establishment of steam communication with the Far East and the Antipodes will to some extent make clear the problem which the famous *Great Eastern* was intended to solve. So far, the largest vessels dealt with have been the *Persia* and the *Himalaya*. In them was seen the results of a gradual development over a period of twenty or thirty years; in the *Great Eastern* was witnessed an audacious experiment on an unheard-of scale. Constructed for the Eastern Steam Navigation Company, with the object of steaming to Ceylon and back without re-coaling, she was the outcome of the ideas and collaboration of Brunel and Scott Russell, and was built at Millwall during the years 1854 to 1857, and, after much difficulty, was launched on January 31, 1858.

Her principal dimensions were as follows:

Length, 692 ft.
Breadth, 82½ ft.
Depth, 58 ft.
Power of screw engines, 4886 indicated horse-power.
Power of paddle engines, 3411 indicated horse-power.
Coal capacity, 10,000 tons.
Displacement, 27,384 tons.
Speed, 14·5 knots.

A comparison of these figures with those of the *Himalaya* and *Persia* will show that she was five or six times larger than those vessels. Her size, her capacity and power were, however, not more remarkable than her strength. As a ship she was the wonder of her day; as an example of ship construction she has been praised by the greatest naval architects; as an achievement she was the product of an age alive with optimistic and grandiose ideas. In 1851 the Great Exhibition in Hyde Park had displayed to an astonished world the wonders of science and engineering. In twenty years England had been covered by networks of railways and telegraphs. Steamships were finding their way into every ocean, and there appeared to be no limit to the conquests of the engineer possible on land and sea. Of funds for any great scheme, there seemed to be ample, as the railway mania had shown, and this idea of untold wealth was not lessened by the stories from the newly found gold fields of California and Australia. Of great projects there were many, and that of Brunel for a gigantic ship was not a whit more extraordinary than that of De Lesseps for cutting a canal from the Mediterranean to the Red Sea or that of Brett for connecting Europe and America by a submerged electric cable. As it turned out, the last of these schemes owed some of its final success to the *Great*

*Eastern,* which, in 1865, took aboard the 3000 miles of cable for connecting Valentia and Newfoundland. It is true that as a commercial venture the *Great Eastern* was a complete failure; it is true she never carried a passenger to Australia and that her voyages across the Atlantic were attended with little success, but had marine engineering been in a more advanced state the tale might well have been different.

To deal adequately with her construction and machinery is here impossible, but one or two matters of general engineering interest may be recalled. For some of the structural ideas embodied in the *Great Eastern,* the builders, Scott Russell and Company, were indebted to Robert Stephenson and Fairbairn, and by a coincidence the ship was built on the site on which Fairbairn carried out the large-scale experiments for the Menai Straits Tubular Bridge. With her complete cellular double bottom, her cellular double upper deck and her strong longitudinal bulkheads, the ship—as a transverse section shows (see Fig. 8)—was something like the bridge, and she was equally fit to withstand the stresses to which she was subjected. Before the building of the *Great Eastern,* most of the ideas of the iron ship-builder were obtained from the constructors of wooden ships. The *Great Eastern,* however, was an entirely new departure and one feature, the double bottom, proved her salvation. The *Great Britain* had shown how an iron ship could survive stranding; the *Sarah Sands* how an iron ship could escape destruction from fire; while the *Persia's* collision with an iceberg taught yet another lesson. The *Great Eastern* in her turn was to make another record, for in August 1862, she struck a submerged rock off Long Island, New York, ripping a hole in her outer skin, yet without the water finding its way into the ship. The useful career of this noble ship came to a close in 1873. She had laid the Atlantic cables of 1865 and 1866, she had laid the French cable to America in 1869, the Suez to Bombay cable in 1870 and another Atlantic cable in 1873, after which no further use was found for her and she fell into disrepair. She changed hands several times, and in November

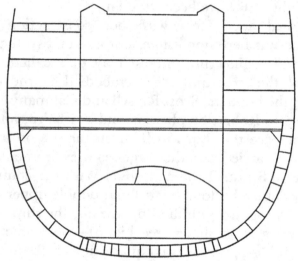

Fig. 8.  Cross-section through *Great Eastern*
By courtesy of the Science Museum, South Kensington

1888 she was sold for scrap for the sum of £43,000. The breaking up, it was said, will commence on January 1, and "thus her long career as a money-losing concern" will be terminated.  It was but a poor epitaph for a ship of which David Pollock wrote in 1884, that she "undoubtedly furnished, in large measure, the experience that has recently been causing so great a change in the tonnage of our mercantile marine", and "apart from commercial considerations, however, this premier leviathan stands out as a wonder and pattern of naval construction".

PLATE V

*GREAT EASTERN* UNDER WAY, JULY 23, 1865

From Russell's *The Atlantic Telegraph*

## CHAPTER VIII

# EARLY IRON WARSHIPS

THE use of iron for the construction of fighting ships involved considerations other than those which had to be taken into account in the design of merchant ships, and it was not till after the Russian War of 1854–6 that the iron-hulled iron-armoured steam man-of-war was seriously considered as a possible rival of the wooden line-of-battleship. The relative progress of the use of iron in both the British mercantile marine and the Royal Navy is shown by the figures quoted by Sir William White in his *Manual of Naval Architecture*, 1877. Of shipping, both sail and steam added to the British mercantile marine in 1850, the proportion of iron ships was one-tenth: in 1860, one-third; in 1868 nearly one-half. The iron steam ship, however, progressed more rapidly than the iron sailing ship, the proportion of iron steam ships in the Register in 1850 being one-fifth; in 1860 five-sixths; and in 1868 nine-tenths. In the Royal Navy the proportion of iron ships to wooden ships in 1850 was less than one-fifth, in 1860 it had fallen to less than one-tenth; but by 1870 it had risen to one-third, and when White published his *Manual*, not a single wooden ship was being constructed for the Navy.

Our modern Navy may be said to have had its birth when the *Warrior* and *Black Prince* were built. With iron hulls and iron armour these vessels mark clearly the supersession of the wooden ship of the line by the modern battleship in its earliest form, spoken of by Engineer-Commander F. L. Robertson in his book,

*The Evolution of Naval Armament*, 1921, as "the most dramatic, swift and far-reaching change which has ever befallen war material". With the remarkable revolution which took place in the 'sixties of last century we are not here concerned, the object of this chapter being to trace the introduction of the iron fighting ship, to review the circumstances which led to the temporary abandonment of the use of iron, and to recall the history of some of our early iron men-of-war.

Among other matters dealt with by Robertson is the work of the famous French artillerist General Paixhans (1783–1854), to whom we owe many innovations, and who in his work, *Nouvelle Force Maritime et Artillerie*, 1822, suggested the possibility of rendering ships invulnerable by casing their sides with iron plates. The idea of using iron for armouring vessels was by no means new and Brassey quotes the instance of the floating batteries of Chevalier D'Arcon, used in the attack on Gibraltar in September 1782, which had their sides protected with bars of iron. After Paixhans, Captain Ross in his work on *Navigation by Steam*, 1828, hints at the desirability of using iron for making steam vessels for coast defence proof against shot. But these suggestions for iron armour bore little fruit until after the experiences in the Black Sea and Baltic nearly thirty years after Ross wrote, and all the earliest iron vessels built for naval purposes were unarmoured.

The use of iron ships in actual warfare begins with the interesting story of the *Nemesis* (see Pl. V), a vessel built under some degree of secrecy at Birkenhead by Laird for the East India Company. A description of the vessel was given in the *United Service Journal* for May 1840, but of her voyage round the Cape to India, and of her useful work in the China War a full account is given in *The Narrative of the Voyages and Services of the Nemesis from* 1840 *to* 1843, published in 1844 by W. D. Bernard. Laid down in 1839, and completed in March 1840, the *Nemesis* was about 700 tons burden, being 184 ft. long, 29 ft. wide and 11 ft. deep. Her engines of 120 horse-power were made by Forrester, of the Vauxhall Foundry, Liverpool. The ship was divided into water-tight compartments by six iron bulkheads, on each of which was a cock for draining the next compartment if necessary. Designed very flat in the floor, and without a fixed keel, she was fitted with two centre-board lifting keels, 7 ft. by 5 ft., one forward and one abaft the engine spaces. From the first she was commanded by William Hutcheon Hall (1797–1878), who by his experience in the China Seas, and by his knowledge of steam gained in Glasgow and in service in steamers in the United States, was admirably fitted for the post. Hall's first officer was Lieut. Pedder, whose name is re-called to-day by a wharf in Hong Kong, and Bernard gives the names of six engineers who served in the ship, one of whom, Colin M'Dougal, was apparently killed in action. Leaving Liverpool for Portsmouth in March 1840, errors of the compass led to the ship grounding on The Stones in the bay of St Ives. Repaired at Portsmouth she cleared for Odessa, only Hall knowing her true destination, and, steaming at 7 or 8 knots, she arrived at Madeira on April 6. From thence she proceeded to Prince's Island, where she took in some 70 tons of hard wood for fuel. She also called at the island of St Thomas, where the *Enterprize* had revictualled in 1825, and, proceeding in much the same fashion as the *Enterprize*, sailing and steaming, she reached Table Bay in July. Continuing on her voyage, she safely passed Danger Point, where twelve years later the *Birkenhead* was wrecked, but running into heavy weather, she was nearly overwhelmed. On July 17, the sea smashed one of the paddle wheels, and the next day vertical cracks 2 ft. long were discovered in her side plating just forward of her sponsons. Temporarily strengthened with planks fastened to the deck, and with struts and bolts between her angle-iron ribs, she reached Delagoa Bay, by which time the cracks had extended another 5 ft. Further strengthened and repaired, she was able to proceed to Mozambique, and six months after leaving England she reached Ceylon, being the first iron ship to make such a voyage. From Ceylon she proceeded to China, where during the operations against Canton and other cities, she took a prominent part. Her light draught, her handiness and the skill with which she was handled made her one of the best known ships in the fleet, earning for her captain the name "Nemesis Hall". She was the first ship to pass the Boque Forts— the Dardanelles of China; she surveyed the tributaries of the Canton River, and afterwards took part in the actions against Amoy, Ningpo and Woosung, which led to the opening of the Treaty Ports. During the campaign the fleet under Admiral Sir William Parker had been joined by other steam vessels, among them the iron steamers *Proserpine*, *Phlegethon*, *Pluto* and *Medusa*, all built for the East India Company, to which, therefore, belongs the credit of first using iron vessels for war purposes. Bernard's account of the *Nemesis* concerns the period March 1840 to January 1843, when the ship had been sent back to Calcutta, and in his concluding remarks he says:

Sufficient evidence has been given of the utility of iron steamers of *moderate* size, in service upon an enemy's coast. The danger which some have apprehended from the rusting of the rivets by which the iron plates are fastened together, or from their *starting*, through the concussions to which the vessel may be liable, was proved to be almost totally unfounded.

From Calcutta the ship was sent to Bombay to refit, and an official wrote from this place:

The *Nemesis* has been for some time past in our docks, and I have carefully examined her. She displays in no small degree the advantages of iron. Her bottom bears the marks of having been repeatedly ashore; the plates are deeply indented in several places, in one or two to the extent of several inches. She has evidently been in contact with sharp rocks, and one part of her keel-plate is bent sharp up, in such a way that I could not believe that *cold* iron could bear; indeed, unless the iron had been extremely good, I am sure it would not have withstood it without injury. Her bottom is not nearly so much corroded as I expected to have found it, and she is as tight as a bottle.

The *Nemesis* and the other iron steamers, as has been said, were the property of the East India Company, and thus, though they formed a part of the British Squadron in China, their names do not appear in the Navy List. But the year the *Nemesis* sailed, Laird built the *Dover*, of 224 tons and 90 horse-power, and it is with this vessel that the history of the iron ship in the Royal Navy begins. In the same year Laird also built the iron steam vessels *Albert* and *Wilberforce*, each of 459 tons, and the *Soudan*, of 250 tons, all of which were added to the Navy and were sent in 1841 by the Admiralty to explore the River Niger. Though the expedition failed through the spread of sickness, yellow fever being then rife on the coast, the *Albert* and *Wilberforce* did good work for many years in African waters.

A further step in the adoption of iron was taken in

1843, when Robert Napier, who, immediately after supplying the engines to the first ships of the Cunard Company, had added to his activities by opening a yard at Govan for the building of iron ships, was given the order for the three ships H.M.SS. *Bloodhound*, *Jackal* and *Lizard*, vessels about 150 ft. long and 360 tons, with engines of 150 horse-power.

It is stated in the *Life of Napier* that "these were the first iron vessels in the service", but it will be seen that this was not the case. One of them, however, was the first iron vessel in the Royal Navy employed in actual warfare, and about the same time that they were ordered, steps were taken not only to add several iron vessels to the Navy, but also to build iron steam frigates. This action was not taken without raising considerable opposition, and in the House of Commons, in 1843, Captain Charles Napier, then Member of Parliament for Marylebone, remonstrated with the Admiralty, and spoke of the absurdity of building five or six iron steamers without previously trying one. In the reply to his remarks, it was said that no fewer than 40 were to be built; but this was never done. By this time evidence had been forthcoming that iron shot passing through iron plates led to great showers of splinters, and some authorities therefore did their utmost to prevent the iron ships being built. Further experiments which were carried out tended to confirm the wisdom of this action. A good deal of information on the question is given in General Sir Howard Douglas's *A Treatise of Naval Gunnery*, fourth edition, 1855, Section VIII, Part II of which is devoted to "Penetration of Shot into Materials". Among the trials he refers to are those made by the French at Metz in 1835, when plates 1·4 in., 1·7 in. and 3 in. thick were fired at from 12-pounder and 24-pounder guns; those made at Woolwich in 1840 to determine the effect of shot upon an iron target lined with a composition of caoutchouc and cork, the iron being $\frac{5}{8}$ in. thick and the composition 9 in. thick; and further trials made at Portsmouth during 1849–51, under the direction of Captain Chads. In these trials at Portsmouth a large iron target was used, representing a section of the iron steam frigate *Simoom*, the target in some trials being filled in solid between the ribs with oak. Summing up the results of the experiments, Douglas said that

thus, it appears that the destructive effects of the impacts of shot on iron cannot be prevented. If the iron sides are of a thickness required to give adequate strength to the ship ($\frac{5}{8}$, or at least $\frac{4}{8}$ of an inch), the shot will be broken by the impact; if the iron plates be thin enough to let the shot pass into the ship without breaking, the vessel will be deficient in strength,

and the conclusion arrived at was that "iron vessels, however convenient and advantageous in other respects, are utterly unfit for the purposes of war". In the 'fifties, other experiments were carried out in America, France and England, in which several layers of iron plating or thicker plates were used and these led to the adoption of iron armour, but the foregoing results of the trials on plates of a thickness used in shipbuilding will assist in understanding why iron, after being introduced, was for a time abandoned for naval vessels.

An examination of the *New Navy List* for July 1846, compiled by Joseph Allen, R.N., shows that out of a total of 576 vessels on the effective list, the iron steam vessels listed in Tables VIII and IX, given on page 119, had been completed or were building.

Of the vessels in Table VIII, the *Dover*, *Onyx*, *Princess Alice*, and *Violet* were employed on the Mail Packet Service at Dover, the *Mohawk* was used on the Great Lakes of North America, the *Fairy*, a favourite with Queen Victoria, was attached to the Royal yacht, *Victoria and Albert*, the *Bloodhound* was in the Mediterranean, while the *Harpy* and *Lizard* were on the southeast coast of America, the *Lizard* taking part in the expedition up the Parana.

While Table VIII contains the names of the first iron vessels in the Royal Navy, greater interest lies with the history of the ships in Table IX, inasmuch as the names include those of the four vessels *Birkenhead*, *Megaera*, *Simoom* and *Vulcan*, which were laid down as steam frigates but which, owing to the results of firing trials, were reduced to the status of troopships. It will be noticed that none of the vessels was built in the Royal Dockyards, in which, indeed, no iron shipbuilding was undertaken till the *Achilles* was laid down at Chatham, while the names of the builders include the firm Ditchburn and Mare, the founders of what was afterwards widely known as the Thames Iron Works and Shipbuilding Company of Blackwall. Three other iron steam vessels, the *Caradoc*, of 676 tons, the *Undine*, of 284 tons, and the *Oberon*, of 649 tons, were added to the Navy shortly after the vessels shown in the tables, after which no more iron vessels were built or purchased till the Russian War.

Of the four steam frigates referred to, the *Birkenhead* (see Pl. V) was the first to be built. Laid down in 1843, she was launched December 30, 1845, and in November 1846 was commissioned for particular service. Employed from time to time, it was while she was on service that the decision was taken to remove

PLATE VI

*NEMESIS*
From Bernard's *Voyages and Services of the Nemesis*, 1844

H.M.S. *BIRKENHEAD*, 1845
From painting by A. W. J. Burgess, R.I.  Courtesy of Cammell, Laird and Co. Ltd.

the armament from the main deck, place a few light guns on the upper deck, where the bulwarks were of timber, and to use her as a transport. In this capacity, she sailed on her ill-fated voyage of 1852. Leaving

### Table VIII. Iron Ships in Navy, 1846—Completed
### (S = Screw ships)

| Name | Tons | Builder | N.H.P. | Engine maker |
|------|------|---------|--------|--------------|
| *Albert* | 456 | Laird | 70 | |
| *Bloodhound* | 370 | Napier | 150 | Napier |
| *Dover* | 224 | Laird | 90 | Fawcett |
| *Dwarf* (S) | 164 | Ditchburn & Mare | 90 | Rennie |
| *Fairy* (S) | 312 | Ditchburn & Mare | 128 | Penn |
| *Grappler* | 557 | Fairbairn | 220 | |
| *Harpy* | 343 | Ditchburn & Mare | 200 | Napier |
| *Jackal* | 346 | Napier | 150 | Napier |
| *Lizard* | 340 | Napier | 150 | Napier |
| *Mohawk* | 174 | — | 60 | |
| *Myrmidon* | 374 | Ditchburn & Mare | 150 | Boulton, Watt & Co. |
| *Onyx* | 292 | Ditchburn & Mare | 120 | |
| *Princess Alice* | 270 | Ditchburn & Mare | 120 | Maudslays |
| *Rocket* | 70 | — | — | |
| *Torch* | 340 | Ditchburn & Mare | 120 | Seaward |
| *Violet* | 292 | Ditchburn & Mare | 120 | |

### Table IX. Iron Ships in Navy, 1846—Building
### (S = Screw ships)

| Name | Tons | Builder | | Engine maker |
|------|------|---------|--|--------------|
| *Antelope* | 649 | Ditchburn & Mare | 646 I.H.P. | Penn |
| *Birkenhead* | 1400 | Laird | 500 N.H.P. | Forrester |
| *Megaera* (S) | 1395 | Fairbairn | 350 N.H.P. | Rennie |
| *Minx* (S) | 301 | Ditchburn & Mare | 10 N.H.P. | Seaward |
| *Sharpshooter* (S) | 503 | Ditchburn & Mare | 348 I.H.P. | Ravenhill |
| *Simoom* (S) | 1980 | Napier | 1242 I.H.P. | Portsmouth |
| *Trident* | 848 | Ditchburn & Mare | 350 N.H.P. | Seaward |
| *Triton* | 650 | Ditchburn & Mare | 260 N.H.P. | Boulton, Watt & Co. |
| *Vulcan* (S) | 1764 | Mare | 400 N.H.P. | Maudslays |

Queenstown in January of that year with large detachments of troops, she had aboard some 638 persons, including a crew of 132. Six weeks later, on February 27, she struck a pinnacle rock off Point Danger, Algoa Bay, Cape of Good Hope, broke in two and less than 200 of those aboard were saved.

Many accounts of the wreck have been given and incidents connected with it have been enshrined in poetry. The disaster had the effect of strengthening the opposition to the use of iron in the Navy. Though perhaps the ship would have foundered in the end, the rapidity with which she went to pieces was hastened through the presence of holes in the bulkheads. Referring to this, Sir William White remarked, "the ill-fated troopship *Birkenhead* is a case wherein the original subdivision was satisfactory, but was marred by cutting openings in the partitions in order to make more easy the passage from compartment to compartment in the hold". The end of this fine ship was the sale at the Cape on February 28, 1853, of portions of her hull, masts, yards and stores for £164. 13s. 6d.

The *Birkenhead* was driven by paddle wheels, having been laid down before the decision had been taken to adopt the screw for naval vessels, but the *Megaera*, *Simoom* and *Vulcan* were all screw ships. The *Megaera* was one of the last vessels built by Fairbairn before he abandoned iron ship building and closed his Millwall yard, and she was on the stocks at the same time that he carried out the experiments on a 75-ft. model of the Britannia Bridge. The ship was 1395 tons by measurement and 1550 tons by displacement, while the engines by Rennie developed 925 horse-power. During the 'fifties, like the *Simoom* and *Vulcan*, she did good work as a troopship and during the Russian War was attached to the Mediterranean Fleet. After many years service, in February 1871, she sailed for Australia under Captain A. T. Thrupp. She had been somewhat neglected, however, and in the Indian Ocean sprang a leak. Faced with the possibility of a disaster, Captain Thrupp decided to beach her on St Paul's Island while yet the opportunity offered, and this was done on June 19. Huts were erected on shore, stores were landed and the ship was abandoned. A month later Lieutenant Jones was taken aboard a Dutch man-of-war; in August the *Oberon* arrived with provisions, and finally the crew was taken off in the *Malacca*. The captain was tried but acquitted. It was shown that the ship had been pronounced unfit for service as early as 1867, and the Government Commission which enquired into her loss censured the Controller of the Navy and other officials.

The third of the frigates, the *Simoom*, was built by Robert Napier, being laid down on December 20, 1845, launched on May 24, 1849, and finished about two years later. The delay in her completion was not due to him, but to the change in the official attitude to the use of iron, and Napier was thereby placed at some inconvenience. It was a replica of a part of the *Simoom's* structure that was used for the firing trials at Portsmouth. Of about 2000 tons displacement, the *Simoom* was 246 ft. long, $41\frac{1}{2}$ ft. wide, and originally had horizontal oscillating engines by Boulton, Watt and Co. of 539 indicated horse-power, giving her a speed of only $8\frac{3}{4}$ knots.

These engines not proving satisfactory, a new set was made at Portsmouth Dockyard, this being the first instance of the construction of propelling machinery by the Admiralty. The new engine had two horizontal cylinders, $62\frac{1}{2}$ in. diameter, with a stroke of 3 ft. In a return of the screw steamers, compiled by the Steam Department of the Admiralty about 1856, the following data are given of the trials of the *Megaera* and the *Simoom*:

|  | Dis-placement | Speed | Steam pressure | Revolu-tions | I.H.P. |
|---|---|---|---|---|---|
| *Megaera* | 1554 tons | 10·24 | 8 lb. | 74 | 925 |
| *Simoom* | 2830 tons | 10·64 | 20 lb. | 55 | 1242 |

The *Simoom* was paid off at Devonport, January 1879, and in 1887 was sold out of the Service.

Each of the three vessels just dealt with was constructed by engineers who had taken up iron shipbuilding. The *Vulcan*, on the other hand, came from a firm of shipbuilders, destined to rank high among their contemporaries. Thames-built ships had long borne a high reputation, and a Hull shipbuilder had stated to a committee of merchants and ship-owners that "ships built on the Thames are unquestionably better than those built at the outports". This was said before Ditchburn and Mare had started in business, but for many years no finer ships were constructed than those of Blackwall, where the *Vulcan* was laid down. Thomas Joseph Ditchburn, who was born in 1801, and died in 1870, had been a shipwright apprentice at Chatham Dockyard and had been employed under the famous constructor, Sir Robert Seppings. Leaving the Admiralty service he became manager to Fletcher and Fearnall of Limehouse, and then, becoming acquainted with Charles John Mare, joined with him in opening a yard at Deptford. Unlike Ditchburn, Mare knew little of shipbuilding. Born in Staffordshire, in 1815, he had been placed with a firm of solicitors in Doctors' Commons, but finding the work uncongenial he turned to other pursuits. His father dying, he leased the family house in Cheshire, and it was with his capital that the partners began work. A fire having destroyed the Deptford yard, another was opened on the east side of Bow Creek at Blackwall, and it was there the early iron vessels they built for the Navy were laid down. Ditchburn retiring in 1846, Mare then added to the establishment by extending the works to the west side of Bow Creek, and the first vessel laid down in the new yard was H.M.S. *Vulcan*. From the same slip in after years came the *Himalaya*, the *Warrior*, the *Blenheim*, and many other famous vessels for both mercantile and fighting fleets. Mare's control of the works, which it was said employed 3000 or 4000 men, came to an end in 1856. After constructing the *Himalaya*, he apparently took a contract at a low price for a number of gun boats, became insolvent, and the business was taken over by his father-in-law, Peter Rolt, and in 1857 was formed into the Thames Iron Works and Shipbuilding Company. After leaving Blackwall, Mare was associated with the formation of the Millwall Iron and Shipbuilding Works, where no less than £100,000 were spent on a rolling mill for iron plates and armour, but this company only lasted till 1866, when it came to an end through the failure to launch H.M.S. *Northumberland* and the disastrous failure of the bankers Overend, Gurney and Company. The history of Mare during his later years is unknown, save that he died in comparative obscurity, on February 9, 1898. A short time after his death a memorial to him was placed in the entrance hall of the West Ham Municipal College, and on this is a long inscription recalling the work of Mare, and also a half block model of the *Vulcan*. Launched in 1849, the *Vulcan* was 220 ft. long, 41 ft. in breadth, had a draught of 17·9 ft. and a displacement of 2396 tons. Her original engines, by Rennies, had four cylinders, 42½ in. diameter, with 2-ft. stroke, which, supplied with steam at 8 lb., developed 792 horse-power. These were afterwards replaced by a set of engines by Maudslays of slightly greater power. After some years' service as a troopship, the *Vulcan* was sold out of the Service in 1867 and converted to a sailing ship.

Obtained from many sources, these notes on our early iron warships have been considered worthy of inclusion in a review of the history of naval and marine engineering, as they recall the second great incursion of the mechanical engineer into the sphere of naval construction. The experiences of the 'forties did not lead at once to the extended use of iron in most navies, but they prepared the way. It was the urgent necessity, which arose during the Black Sea campaign, for protecting ships from shell fire which led the French to adopt iron armour, and their three wooden iron-encased batteries *Dévastation*, *Tonnante* and *Lave*, successfully took part in the attack on Kinburn, October 17, 1855. These vessels were followed first by our own iron-hulled iron-armoured batteries *Erebus* and *Terror*; then, by the reconstructed French line of battleship *Gloire*, a wooden vessel armoured from stem to stern; but a much fuller application of the art of the mechanical engineer to the sea-going fighting ship was seen in the *Warrior* and *Black Prince*, iron vessels of 9210 tons displacement, protected by 4½ in. of iron armour and driven by engines of 5470 indicated horse-power, built, respectively, at Blackwall and Glasgow in 1861.

## CHAPTER IX

# LOW PRESSURE MARINE BOILERS

THE outstanding feature in marine engineering practice during the first fifty years of the steam ship was the adherence to the use of steam at low pressure. Engines increased greatly in size and power, and there were such notable innovations as the introduction of tubular boilers and direct-acting engines, but low-pressure steam, salt-water feed and jet condensers were almost invariably the rule. There was an immense variety of designs of boilers and engines, yet the maxims which sufficed for the engineers of the 'thirties met the needs of the engineers of the 'fifties and early 'sixties. That this was the case can be seen by a comparison of Robert Murray's *Rudimentary Treatise on Marine Engines and Steam Vessels*, third edition, 1858, with the works of Otway, Robinson, Gordon, Williams and others. Murray was an engineer surveyor of the Board of Trade and therefore conversant with mercantile practice, as the other writers had been familiar with naval practice. He wrote at a time when there were still doubts as to the superiority of the screw and as to the capacity of steam vessels to make long voyages, and his book is a valuable source of information regarding the state of marine engineering at the time of the Russian War and the building of the *Great Eastern*. To Murray we are also indebted for the earliest engineering papers contained in the *Transactions of the Institution of Naval Architects*. In one of these, he gave particulars of fifty merchant steamers sailing from Southampton, Portsmouth and Weymouth, from which further information on the position of marine engineering practice in 1860 can be obtained.

Nearly all early marine boilers were large box-like structures with internal square furnaces and long narrow passage-like flues. The sketches in Figs. 9*a* and 9*b*

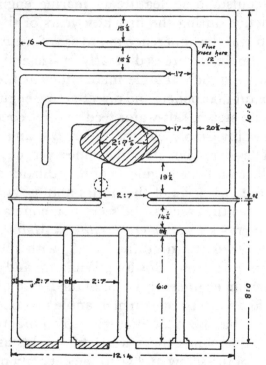

Fig. 9*a*.  Horizontal section of flue boiler
for the *Comet*, 1828

show the construction of a copper boiler made for H.M.S. *Comet* by Maudslays in 1828. A typical example of a naval boiler installation was that in H.M.S. *Driver*, 1060 tons, 280 horse-power, engined by Seaward in 1841. There were three separate boilers, each 26 ft. long, 9 ft. wide, and 12½ ft. high, each boiler having three furnaces, 7 ft. 8 in. long, 2 ft. 6 in. wide, with flues 18 in. wide, and water spaces 5 in. wide. The circuit of the gases from furnace to funnel was no less

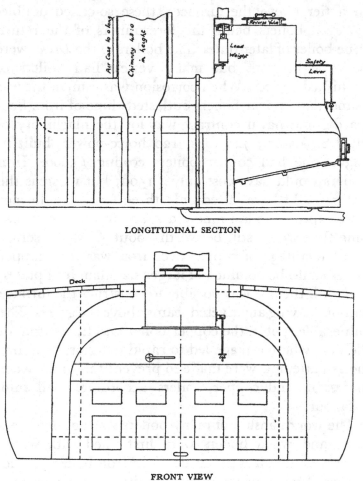

LONGITUDINAL SECTION

FRONT VIEW

Fig. 9*b*. Longitudinal section and front view of
flue boiler for the *Comet*, 1828

Boulton and Watt Collection, by courtesy of Birmingham Reference Library

than 70 ft. The weight of the boilers was about 40 tons and at working height they contained about 50 tons of water. The working pressure was $4\frac{1}{2}$ lb. per square inch.

Following the usual practice, the furnace and flues were on the same level, but in some vessels the flues were in a tier above the furnaces, these so-called double-storeyed boilers being the forerunners of the return tube boiler of later times. The boilers of the *Driver* were of wrought iron, but many vessels had boilers of planished copper. When corrosion was rampant it was found that copper boilers outlasted those of iron. Seaward's first naval contract was for the machinery of H.M.S. *Volcano*, 720 tons, 120 horse-power, built in 1836. She had copper boilers costing £5000. Iron boilers would have cost only £1500; but whereas the latter would have lasted only three years, the former were expected to last nine years, and at the end of that time they would still be worth about £3000 as scrap. One advantage of copper over iron was that copper plates could be obtained of larger size than iron plates, and with them it was possible to construct the furnace without having any riveted seams above the grates. The vulnerable spot in the copper boiler was the bottom of the flue, where leakage led to rapid deterioration, and though attempts were made to prevent this, none were successful, and copper boilers accordingly fell into disfavour.

The workmanship of many boilers was not of a high class, and when boilers were first filled with water there were innumerable leaks. "In iron boilers", said Dinnen, "when water is first admitted after construction, hundreds of 'weeps', or channels in the plates and rivets, whence water oozes, are totally disregarded, the most important only being stopped mechanically, the rest are staunched merely by the rust the water has formed in its passage." Writing from Woolwich in 1839, Peter Ewart instructed Dinnen at Chatham to have two sacks of oatmeal scattered uniformly down the water spaces of the boilers of H.M.S. *Hecla*, before the water was put in. Dung, potatoes and other substances were also used to prevent leakage. When placed aboard, boilers were supported on a platform of planks fixed to oak bearers, the boilers being bedded on a layer of mastic cement consisting of sand, powdered stone and glass, litharge, grey oxide of lead and oil, which rapidly hardened. Owing to this method of bedding, it was not an unknown thing for the boilers to remain usable although the bottom plating was corroded through. The boilers altered in shape considerably under steam, and were even less suitable for withstanding external pressure than internal pressure. One of the weakest parts of the boiler was the flat bottom of the furnace which was subject to an upward pressure equal to the steam pressure plus the head of water in the boiler. In the long winding flues soot collected to a great depth and as the flues were only 16 to 18 in. wide, the cleaning, which, according to Naval Regulations, had to be done after the fires had been alight 144 hours, was a tedious and disagreeable task.

With very few exceptions salt water was used in the boilers, and the prevention of serious incrustation was one of the principal duties of the engineer. The density of the water was tested by drawing off a small quantity into a saucepan, bringing it to the boil, and then taking the temperature with a "common brewer's thermometer". If the temperature exceeded 215° F. there was danger of incrustation. Hydrometers were supplied later on, but these were graduated for a temperature of 55° F. and were found of no use in hot climates, where it was impossible to cool the water to that temperature. Blowing down was carried out every two hours, this

practice superseding the old plan of completely empty-
ing the boiler every third or fourth day. Maudslays
introduced brine pumps and valves and Seaward brine
tanks, but these were found to be unnecessary com-
plications. If methodically carried out, the blowing
down ensured the boilers being kept remarkably clean,
and its importance was fully recognised. "I have considered blowing off", wrote Dinnen, "at all times, so
important and dangerous an operation, that I have
never permitted it to be per-
formed without the attendance
of one fireman at least, with
the engineer of the watch,
whose undivided attention was
directed to trying the gauges,
till the moment prescribed for
shutting the cocks". The invention of the "Kingston" valve
(see Fig. 10) in the bottom of
the ship and of the safety spanner
which could not be taken off
the blow-down cock while the
cock was still open was the work

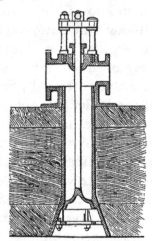

Fig. 10. Kingston valve

of John Kingston, of Woolwich Dockyard, who in 1837
was awarded a medal by the Society of Arts for his valve.
Sometimes blow-down cocks jammed. In this case the
valves of the feed pump would be reversed and the
brine pumped out. In an extreme case, says Murray,
"a usual plan is to knock out a rivet from the bottom
of the boiler". In the naval steamers running mails
from Falmouth to the Mediterranean, some of the vessels had to have a general repair of their boilers every
fourth voyage. Neglect with regard to blowing down
might lead to the water spaces becoming solid with

salt. The boiler-feed water was obtained from the discharge from the air pump and with very low pressures the boiler was fed by gravity from a tank above the upper deck. An annular casing around the base of the funnel sometimes served the same purpose as the modern economiser. Originally test cocks were fitted for determining the height of the water in the boiler.

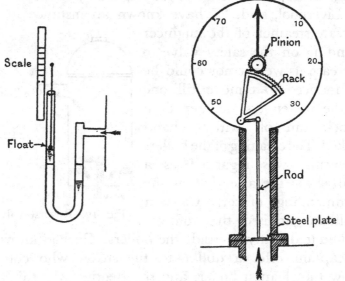

Fig. 11. Early pressure gauges

Then came various forms of mechanical indicators worked by floats, and later on, the glass-tube water-gauge. Mechanical indicators were favoured by some engineers because they could be seen from the deck.

The fittings on the boiler included, beside the water indicator, deadweight safety valves—a type used at sea till the 'seventies—a mercurial pressure gauge consisting of an iron U-tube half filled with mercury, fitted to the front of the boiler (see Fig. 11), blow-down cocks, feed cocks and a reverse or atmospheric valve. The last (see Fig. 12) was a valve which opening inwards would automatically admit air into the boiler when a partial vacuum was formed owing to the cooling down of the water. Accidents arose from many causes, one of which was increasing the load on the safety valve. One engineer in the Navy was discharged for "persistently altering the safety valves contrary to orders", and Captain Chappell, the Superintendent of the Admiralty mail vessels at Liverpool, said, "I have known an instance in a private steamer of the engineer standing on the safety valve to increase its weight nor could he be induced to remove till one of the passengers threatened to knock him off with a hand spike." The cleaning of the boilers internally was regarded as a matter for the attention of the commanding officer. Captain Williams advised that officers

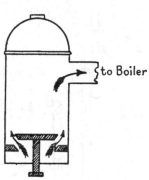

Fig. 12. Reverse valve

should frequently go inside the boilers. He had known a captain offer a dollar to the stoker who could show the cleanest boiler, and so effective was this as an incentive that in the end he had to give a dollar for each boiler. Cleaning of boilers was ordered to be done after steam had been up from eighteen to twenty-four days, and particular attention was to be paid to those parts exposed to the action of the fires. As soon as the boilers were cleaned they were given a coating of tallow, this being preferable to black lead which caused the formation of froth. Boilers were sometimes injured by the use of heavy hammers for removing the scale and by lighting fires of shavings to crack it off.

In spite of their great size and weight, and the great

mass of water they contained, flue boilers were generally fitted in mercantile vessels until the 'sixties,
although the Navy discarded them much earlier. The
first vessels in the Royal Navy to have tubular boilers
were the *Penelope*, 1843, and the *Firebrand*, 1843, the
latter having in her boilers 2250 brass tubes, 2 in. diameter, $\frac{1}{8}$ in. thick and 5 ft. 1 in. long. Lord Dundonald
also fitted tubular boilers in H.M.S. *Janus* in 1845.
Tubular boilers had also been fitted in river craft, the
*Luna*, built by J. and W. Napier in 1837, being the first
vessel on the Clyde to have a tubular boiler. An interesting, and at one time a popular, boiler was that
patented by Lamb and Summers, and used in the P.
and O. vessels in the 'fifties. In this boiler the tubes
were replaced by narrow vertical flat-sided flues, $1\frac{3}{4}$ in.
wide, this modification being adopted on account of the
difficulties of keeping tubes clear of soot and fine ashes.
A description of the boilers and particulars of some
ships in which the type was used were given in a paper
read by Andrew Lamb before the Institution of
Mechanical Engineers in 1852.

The significance of the terms low-pressure steam and
high-pressure steam has changed with each generation.
A hundred years ago Jacob Perkins had constructed
boilers and engines for working at many hundred
pounds' pressure; yet, in marine engineering, the term
high-pressure was applied to steam at even 15 lb. per
square inch. It may be taken that in the 'thirties the
ordinary steam pressure in marine boilers was 5 lb.; in
the 'forties, 10 lb.; and in the 'fifties, 20 lb.; and vessels
using steam at any pressures in excess of these would be
considered as using high-pressure steam. The earliest
recorded instance of the use of so-called high-pressure
steam in the Navy is that given by Otway, commanding

officer of the *Echo*, 294 tons, 140 horse-power, in which
the experiment was made. Originally the *Echo* had
boilers working at $3\frac{1}{2}$ lb. pressure, but through the
efforts of "the late Captain King, of the Packet Establishment at Falmouth, aided by Mr Ward, a very able
practical engineer", the vessel was given boilers working at 15 lb. per square inch. The boilers were cylindrical, with internal cylindrical furnaces and flues.
They were 26 ft. in length. Otway, who himself was
a believer in higher pressures, said that whereas with
$3\frac{1}{2}$ lb. pressure the vessel used 12 to 14 bushels of coal
per hour, with steam at 15 lb. pressure, the consumption was only 7 to 8 bushels. This was an experiment
which, had it been followed up, might well have hastened the progress of marine engineering. That no
similar experiments were made was largely due to
timidity and a want of knowledge of the theory of heat.
The fear of the consequences of the explosion of a high-
pressure boiler on board a ship was a very real one,
while as regards warships there was the additional
dread of the results of the effect of a shot entering a
boiler. With regard to the nature of heat, even in the
fourth edition of Main and Brown's *Marine Steam Engine*
all that the engineer was told, was that, "It will be
sufficient for us to state, that the cause of heat, whatever
it may be, is called caloric." Under the circumstances,
it is not surprising to find that very different views were
held as to the efficacy of high-pressure steam. Captain
Williams, lecturing at Portsmouth, told his listeners
that "with regard to the use of high-pressure steam, it
is known that the mechanical effect is the same under
whatever pressure the steam is raised". Even Joshua
Field declared that the prevalent opinion that high-
pressure steam necessarily meant more power devel-

oped was a mistake, for, he said, "if steam of higher density were generated the quantity was less in proportion as the pressure was increased, and thus the power remained the same". On the other hand, Rennie, who had been asked by the Admiralty to investigate the matter, had reported "that the Cornish system of high-pressure condensing might be applied to the Navy with the greatest advantage". Lieut. Gordon also, in 1845, wrote: "The grand defect of the marine steam engine of the present time is the use of low-pressure boilers instead of high, and it is difficult to understand the prejudice of engineers and practical men against high pressure on board ship, whilst the great advantages of its use on shore can be denied by none."

Field's remarks were made during the discussion of a paper entitled, *On the Employment of High-Pressure Steam working expansively in Marine Engines*, read before the Institution of Civil Engineers in June 1849, by Seaward. The Admiralty had asked several firms for their opinion on the matter, and Seaward had replied: (1) the highest pressure that we have in any case put upon a marine boiler of our own construction was about 16 pounds per square inch; that we can obtain equally good results with steam at a lower pressure. From 10 to 12 lb. is the usual pressure we employ in the merchant service for engines and boilers of comparatively little power. (2) The steam pressure at present employed in the service is about 8 lb. per square inch. We consider steam at that pressure to be well adapted for the exigencies of the service. We believe it is calculated to secure all the important advantages of power, and economy, weight and space in a very economical degree. These advantages will in some respects be slightly increased by augmenting the steam pressure to 10 or 12 lb. per square inch. (3) We strongly recommend that the steam employed in the Navy should not be of greater pressure than 10 lb. per square inch or in extreme cases 12 lb. per square inch. Any material increase in the latter pressure will be attended with considerable risk without any adequate advantage. Besides Field, Farey, A. M. Perkins, Rennie and Scott Russell took part in the discussion, Scott Russell remarking that Seaward's views gave a correct representation of the opinion of the profession, while Rennie, with the lamentable loss of life and property through boiler accidents in America in mind, thought that the attitude of English engineers was "characteristic of that sound practical caution to which they owed much of their reputation"

If in the matter of steam pressure marine engineers proceeded with great caution, many attempts were made by one and another to reduce the steam consumption of engines, and among the experiments of permanent interest were those made with superheated steam. In 1856, trials were carried out in both H.M.S. *Black Eagle* and H.M.S. *Dee* with steam superheated according to the plans of the Hon. John Wethered of the United States. Murray gives an account of the trials, and Wethered read a paper on the subject before the Institution of Civil Engineers in 1860. In these trials a nest of tubes was placed in the base of the funnel and a portion of the steam from the boilers was superheated to 500° or 600° F. This steam was then mixed with the ordinary steam, resulting in a considerable improvement in consumption. The economy of fuel was stated to be 20 per cent to 30 per cent or more, but Murray discussing the results said that "unless we can be assured that the particular boiler which has been made the subject of experiment with Wethered's apparatus is in all respects a good and efficient one according to our present ideas of perfection in a marine boiler, the seeming economy resulting from the process will be more apparent than real". Atherton, Dinnen and Patridge were all concerned on behalf of the Admiralty with the trials, and Patridge declared that in the *Dee*, in which superheated steam had been used for two years, there was not the slightest appearance of any injury to the machinery, nor had there been any deterioration or difficulty with her packing. In 1860, superheaters were also fitted to H.M.S. *Rhadamanthus*, and for a number of years afterwards they were included in most designs for naval machinery.

While it was generally beyond the power of captains and engineers to reduce the steam consumed by the engines of their ships, much attention was paid to the economical use of coal, although the human element often neutralised the efforts that were made. Stokers liked to suit their own convenience, and so long as steam was blowing off from the safety valve who could complain? By the uninitiated, the volume of steam coming from the waste-steam pipe was thought to be a measure of a stoker's capability. Otway would have none of this waste, but what he liked to see was "a gentle lambent vapour playing about the end of the waste-steam pipe". The training of stokers had occupied the mind of Ross as early as 1828, and he laid down that they should be trained to their calling; that they should not be employed on anything but the care of the fire, and that they should be allowed a double quantity of beer or other beverage while the engines were at work. It was recognised that firing should be at short intervals and in small quantities, but stokers proved as difficult to manage as the fires. If stokers did not prove as amenable as officers desired, it was still possible to exercise economy by regulating the coal supply, and by preventing unburnt coal being thrown overboard. The cinders were sometimes mixed with resin and re-burnt. The little *Sirius* on her transatlantic voyage in 1838 used resin, while the *British Queen*, before leaving England for New York always took aboard 50 casks of it. A curious example of stokehold procedure is given by Captain Williams. In March 1841, he tells us, H.M.S. *Stromboli*, on passage from Greece to Malta, steamed at eight knots against a light breeze for seventeen hours with an expenditure of 8 cwt. per hour of bad coals mixed with the following materials:

> Half a cord of wood,
> A cask of mineral tar,
> Three-quarters of a cask of Stockholm tar,
> A cask of pitch and seventy (70) coal bags.

"The cinders were carefully sifted and mixed with pitch; the coal bags were opened out and tar poured over them; they were then rolled up and covered with cinders and small coal; these bags with the addition of a small quantity of wood and coals thrown on occasionally, kept the steam up in three boilers to six and a half pounds upon the square inch." Some commanding officers attempted to lessen waste by creating rivalry between the watches, but others disagreed with this. Again, it was suggested that an effectual check on expenditure could be maintained by locking the bunker doors and weighing out one or more tons at a time.

Only North Country coal was used at first, and there was much trouble from smoke, both from the point of view of the cleanliness of the ship and rigging, and from the point of view of visibility. This was especially the case in the Baltic campaign. With the exception of Napier's flagship, the *Duke of Wellington*, which had Welsh coal, all the ships of the squadron used New-castle coals, "the smoke from which was so intolerable that in coming in, the channels could scarcely be distinguished". The Admiral gave it as his opinion that "in going into action with a fleet, or even with batteries, such a smoke would be injurious to the correct performance of evolutions, and that, therefore, it was a matter of great importance that coals making the least smoke should in future be supplied". Patent fuel had been tried as early as 1839, when the *United Service Magazine* recorded the trip made by some of the Lords of the Admiralty in H.M.S. *Firebrand*. On this trip fuel prepared from screen and refuse coal mixed with coal tar, Stockholm tar and Trinidad pitch was used. It was reported that with this fuel the consumption was less than with the best coal, that it raised steam quicker, that there was less ash and smoke, and that it saved labour and did less injury to the boiler and to the furnace bars. Both the Royal Navy and the Mercantile Marine were afterwards indebted for their information as to the relative value of the various kinds of coal to the geologist, Sir Henry de la Beche and the chemist, Dr Lyon Playfair (afterwards Lord Playfair), who, in 1848–49, published a *Report on the Coals Suited to the Steam Navy*, giving the results of their investigations and trials carried out at the direction and cost of the Admiralty.

The supply of coal in various parts of the world and the methods of coaling ship were among the matters which occupied the attention of many. Ross, when writing on convoys, said that each merchant sailing ship should be bound to carry a certain quantity of coals, and that a part of these should be kept in bags ready to whip into a boat when wanted, or if the ship is in tow to be swung into the steam vessel alongside her. At home, it was usual to send convicts from the hulks to assist in coaling naval vessels, but in 1837, Otway wrote: "Woolwich and Chatham are now, however, the only dockyards infested with these pampered delinquents, whose very movements are characteristic of their moral dispositions, being thieves of time, for their whole day's duty is not worth an hour's purchase." Abroad there were neither fleet colliers nor coaling wharves, and the coal had to be transported in small boats. H.M.S. *Vesuvius*, in September 1842, coaled at Beyrout. She began coaling on Thursday, September 1. On Friday she received 729 bags in 21 trips; on Saturday 451 bags in 13 trips; and on Monday 228 bags in 7 trips, the total coal taken aboard being $155\frac{8}{11}$ tons. Her log records her coaling at Beyrout again on October 22, and four days later is the entry: "11.30 Punished Jno. Cannon, James Wyles, D. McLean (stokers), 30 lashes each. W. Duff (ordy), 18 lashes, and Jno. Denman, marine, 3 class, with 12 lashes, for getting drunk while coaling ship on the 22 inst. as per warrants." A much more satisfactory rate of coaling ship than that of the *Vesuvius* is given by Captain Ryder, who says that the paddle frigate *Sidon* (Capt. Henderson) took aboard 310 tons of coal in nine hours. Coal was sometimes obtained from sailing colliers, and Williams, after explaining how to place a collier alongside, said:

As soon as the two vessels are fairly placed alongside, let every man in the ship except the permanently excused idlers, be upon the stay whips; have a couple of fiddlers going, serve out the extra allowance of grog authorised by the Admiralty, and let the ship's cook have hot fresh water ready for the men to wash themselves in at dinner and supper, and you may thus hoist in at the rate of an hundred tons a day, which soon finishes the matter.

Music, it may be remarked, continued occasionally to enliven the drab duty of refuelling ship in the Navy until oil banished the evolution of coaling from the ship's routine.

## CHAPTER X

## TYPES OF MARINE ENGINES

WHILE none of the early marine boilers have been preserved, a few early engines are still in existence, and in the Science Museum, South Kensington, can be seen a fine collection of models of both paddle engines and screw engines. Many of these are beautiful examples of the model-maker's art, and they recall a period when marine engineering was a flourishing industry on the Thames. The models are well described in the catalogue of the Marine Engineering Collection, in which are also useful notes on the progress of marine engineering. If further information on the subject is desired, no works are more worthy of study than those of John Bourne. In their day Bourne's writings enjoyed a great reputation among engineers, and the reader can but admire the industry displayed in their compilation. Bourne himself was a pioneer in various ways, but it is by his books he is best known. His *Treatise on the Steam Engine*, first published in 1846, had gone through five editions by 1861. His *Treatise on the Screw Propeller* first appeared in 1852. We owe also to him *A Catechism of the Steam Engine* and *Recent Improvements in the Steam Engine*, 1865.

Bourne's books were well illustrated, and in the treatises on the steam engine and the screw-propeller are to be found three plates (1) of direct-acting paddle engines, (2) of geared screw engines, and (3) of direct-acting screw engines. The engines of all the leading makers—Maudslay, Penn, Napier, Miller, Watt, Rennie, Seaward, Tod and Macgregor, Thomson, and of Scott and Sinclair—are all shown, as well as the pendulum engine of Ericsson and the disc engine of Bishopp. The three plates thus form a valuable record of the principal designs of early marine engines. But though Bourne included some forty-five engines, he did not, by any means, exhaust the subject, for there was, indeed, an almost endless variety of designs. Cylinders were sometimes vertical, sometimes horizontal, sometimes inclined upwards, sometimes inclined downwards. One classification of the types of engines divided them into the three groups: (1) those with parallel motion, (2) those with guides, and (3) those with oscillating cylinders, but the briefest examination of the many engines described by Bourne will show that there was little to be gained by attempting to classify them.

By far the most interesting of all paddle-wheel engines, from the historical point of view, was the side-lever engine, which was but a modification of Watt's land engine (see Fig. 13). With a side-lever engine it would be a simple matter to trace the history of Watt's master patents of 1769, 1782 and 1784, which completely revolutionised the practice of steam engineering. No marine engine, it is true, was fitted with an overhead wooden beam, or with the sun and planet motion, or with tappet gear for working the valves, but the mode of action of the side-lever engine was exactly that of the beam engine as Watt left it. It is perhaps scarcely realised, to-day, how strong was the adherence of the early marine engineers to the ideas of Watt. Shortly after Peter Ewart, the first chief engineer of Woolwich Dockyard, died, James Walker, referring to him in his presidential address to the Institution of Civil Engineers, said:

His knowledge of machines and particularly of the steam engine was very intimate. His admiration for

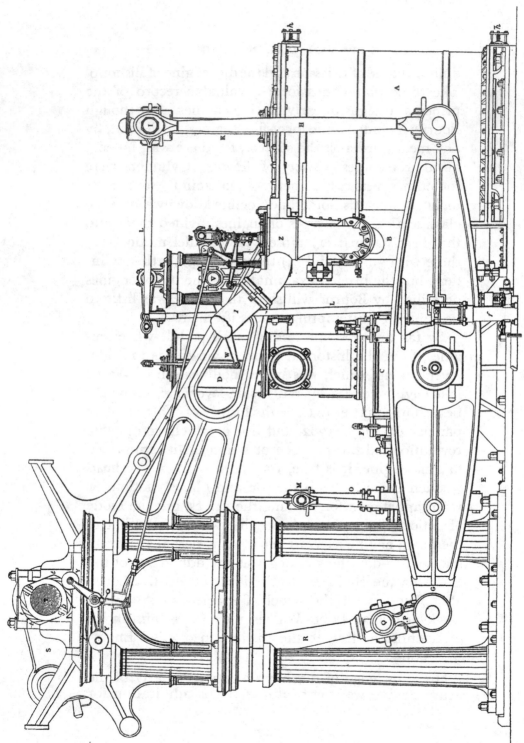

Fig. 13. Side-lever engine

Watt, and his practice at Soho, inclined him to view, with some degree of scepticism, any innovation in the engine which he considered to have been almost perfected by his great master, and for the public situation which he held, this prejudice was probably useful, for the war steamers on active service are not those in which new schemes should first be tried.

That was said in 1843, twenty-four years after the death of Watt. But even before Walker spoke, many marine engineers were breaking away from tradition, and with the introduction of the steeple engine by David Napier, the improvement of the oscillating engine, first by Maudslay and then more particularly by Penn, the invention of the "Gorgon" engine by Seaward, and the twin-cylinder engine by Joseph Maudslay and Field, and the introduction of various forms of direct-acting engines by Rennie, Fairbairn and others, the engine associated with the name of Watt gradually fell into disuse.

All engines for driving paddle wheels necessarily revolved at slow speeds, and when the screw propeller was introduced, the general practice was to adapt one or other of the types of paddle engines, and drive the propeller shaft through gearing, by fitting a large spur wheel with wooden teeth on the engine shaft, and an iron-toothed pinion on the propeller shaft. The *Rattler*, for instance, had a Maudslay twin-cylinder engine driving through such gearing, as did also the *Great Britain*, when re-engined by Penn and fitted with an oscillating engine. Sometimes there were three or four stepped sets of teeth in the spur wheel. Napier, Tod and Macgregor, and other engineers, fitted screw ships with engines with iron overhead beams, also driving through gearing.

During the 'fifties, however, the geared engine was superseded by the direct-acting engine with horizontal cylinders, revolving at higher speed and driving the shaft direct. Reporting on the type of engine most

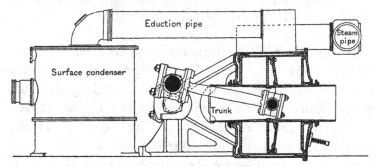

Fig. 14.  Penn's trunk engine

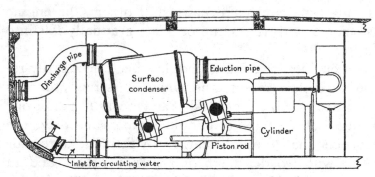

Fig. 15.  Maudslays' double piston-rod engine

suitable for use in the Royal Navy, an Admiralty Committee, in 1858, said

that of all the variety of engines that have been purchased by Government for our screw ships of war, the following are so far superior to all others, that no engines of an older make should ever again be put on board. The engines to which they now refer are:—

(1) The single piston-rod engine, with the connecting-rod attached direct to the crankshaft, and with a single flat guide.

(2) The engine commonly known as the trunk engine, and patented by Messrs Penn and Sons.

(3) The double piston-rod engine.

Sketches of a trunk engine and a double piston-rod engine are shown in Figs. 14 and 15.

Discussing the question further, the committee proceeded to state that after carefully weighing the evidence which had been given, they had "arrived at the conclusion that for very small engines, the single piston-rod engine, as adopted by Messrs Humphrys, Tennant and Dykes is the best engine suited for the Navy". A type of single piston-rod engine, not mentioned by the committee, but which was ultimately to supersede all others, was seen in the vertical inverted-cylinder engines of the *Frankfort*, made by J. and G. Thomson. Bourne gave a sketch of these engines, and said "they are simple, compact and substantial, and upon the whole are a very eligible class of engines for merchant vessels, but for war vessels they, of course, would not be suitable, as they would come above the water line". In general appearance, engines of this type resembled Nasmyth's steam hammer, and they were sometimes called steam-hammer engines. The *Frankfort's* engines were not the first of the kind, for, according to Dr Peter Denny, Caird constructed a set of such engines in 1846 for the *Northman*, a vessel built by Denny for trading between Leith and the Orkneys. A good example of a simple-expansion inverted-cylinder engine was that in the *Laconia* made by J. and G. Thomson (see Fig. 16). This engine had two cylinders 56 in. diameter, 3 ft. stroke.

Just as the general arrangement of the engines differed greatly, so there was no uniformity in details. The framing, the entablatures and bed-plates, the valve motions and other parts, differed with every maker. There was often considerable ornamental work in both

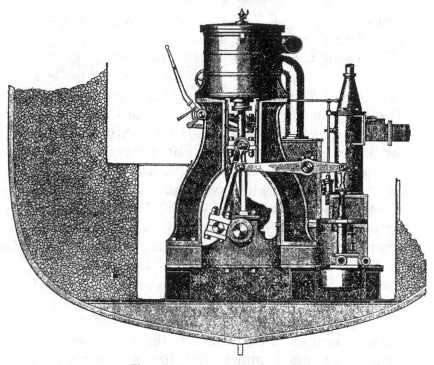

Fig. 16. Engine of *Laconia*, 1856

castings and forgings. Cast-iron columns supporting the entablature carrying the crankshaft in paddle vessels were sometimes decorated like Ionic pillars, or in the Gothic style. Such embellishments, however, soon disappeared. With direct-acting paddle engines, the columns were occasionally of wrought iron, and were braced together much in the same way as the steel columns in destroyers' engines of more recent times.

Cylinders were at first fitted with jackets, to which the steam from the boiler had direct access, and from which it passed into the cylinder through the valve ports. Drains were fitted on these jackets, and in some vessels the supply of drinking water was replenished by the water condensed in them. Otway said that it was found that the jacket of a 40-in. cylinder would yield about 2½ gallons of water per hour, and that in H.M.S. *African*, the use of this water enabled the ship to do without eight of her water tanks. In H.M.S. *Comet*, hot water, when required, was always obtained from the jackets, and Otway suggested that a steam pump should be fitted in vessels for distributing the water throughout the ship. Escape valves on cylinders appear to have been introduced by Maudslays.

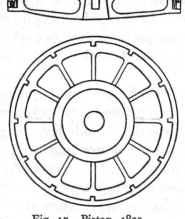

Fig. 17. Piston, 1839

Pistons were at first packed with hempen gasket, although as early as 1791 the Rev. Edmund Cartwright had patented complete metallic packing with springs. The well-known, simple and efficient Ramsbottom ring was not invented till 1854, but before then many forms of metallic packing had been used. Commander Robinson in his book of 1839 gave a sketch of a piston with a ring in circumferential segments behind which were springs (Fig. 17). "The packing rings", he said, "are rings of cast iron which are ground upon one another, and on the ledge of the lower side of the piston, till they fit so accurately upon one another as to be steam tight." The junk ring was also ground tight upon the packing rings. It should be remembered that this was written before Whitworth had brought out his surface plates.

In side-lever and other engines with two cylinders, there was often no common bed-plate, and as the engines rested on wooden sleepers, the cylinders sometimes moved relatively to each other. "It was usual", said Murray, "in well constructed engines to have four horizontal points in an athwartship line on the framing, dressed off so that four points in a true line on the face of a straight edge may lie upon the whole of them and thus prove at any time whether the engines have fallen in towards each other, or fallen away towards the sides of the vessel." As wooden vessels altered shape considerably, the holding-down bolts were slackened off when a vessel was docked or if she went ashore. In the two-cylinder side-lever engine, each cylinder had its own condenser and air pump. The condensers were simply cast-iron boxes. An extension of the condenser and air pump carried the snifting valve through which steam escaped when the cylinder, condenser and air pump were flooded with steam prior to the injection water being admitted.

Before the use of a fixed guide for the crosshead end of the piston rod, various forms of parallel motion were used. In the Naval Regulations of 1837 it was laid down that the first engineer should be able to adjust the length of the various rods and motion, slide valve and eccentric. Such knowledge, however, was not expected from the second engineer. The earliest ship in the Navy to have a fixed guide was H.M.S. *Caradoc*, 1846. The slide valve first used was the long D-valve invented by Murdock. It was driven by a single eccentric, the sheave of which was loose on the shaft, but which was driven round by an accurately placed stop. For starting and reversing, the eccentric rod was thrown out of gear, and the slide valve, which was balanced by a weight, was moved up and down by means of a long lever worked by hand. When the engine had started in the right direction, the eccentric rod would be thrown into gear again. Though Rennie fitted the shifting link, or Stephenson link gear, in the paddle vessel *Sampson* in 1844, the single eccentric arrangement was found in use long afterwards. On one occasion H.M.S. *Archer*, which had single eccentrics, arrived in the Thames after a commission abroad. While on passage, water had been used on the eccentrics to keep the straps cool, and the sheaves had become rusted on the shaft. In this condition the eccentrics were only of use when going ahead, and when the engines were required to go astern the slide valves had to be worked entirely by hand.

With jet condensers, the regulation of the injection water was of equal importance to the working of the steam valves, and care had to be taken that an excess of injection water did not lead to the condenser being flooded. An officer who served in a ship with a simple-expansion engine with jet condenser, said,

Coming into and going out of harbour was a great function, and the watch longest off, of about two petty officers and six men, mustered on the engine-room platform to assist in manipulating the engines, which were hand-starting. Three or four hands were stationed at the starting wheel to move the links up and down, one or two hands attended the main discharge valve, and one to each of the sea injection valves, whilst the engineer of the watch stood by the steam admission and blow through valves and directed operations. This number of hands was not too many, and the question was often asked, "Why are you so long answering the telegraph?", whilst you were doing your best to get the engines started. I quite remember on one occasion the injection valves were left open when the engines were stopped. The whole system filled with water, and before it could be cleared, the telegraph rang to full speed. It seemed an age before the engines started, and when they did, I thought the bottom would be knocked out of the ship.

With very low pressures, steam was admitted to the cylinder right throughout the stroke, but expansive working, as first introduced by Watt, came in at an early date, separate valves being fitted for varying the cut-off in the cylinder. These valves were worked from a series of stepped cams on the crankshaft as shown in Fig. 13. On the maiden voyage of the *Britannia* in 1840, Captain Woodruff's instructions included remarks on working expansively. Various naval writers discussed the matter of expansive working, but it was Commander J. C. Hoseason, who went round the world in the *Inflexible*, who was regarded as the foremost exponent. After commanding the steam vessels *Alecto*, *Prospero* and *Torch*, Hoseason, in 1846, was appointed to the paddle vessel *Inflexible*, 1122 tons, 378 horse-power, engined by Fawcett, Preston and Company. The commission of the ship extended from August 1846, to September 1849. During that time she visited the coasts of China and New Zealand, crossed the Pacific and returned home round Cape Horn, thus circumnavigating the globe and "fulfilling her comprehensive mission in a manner most creditable to her able commander". The paddle sloop *Driver* previously, during a long commission, had circled the globe, but mainly under sail; the *Inflexible's* passages were, however, nearly all done under steam. In the three years she steamed 64,477 and sailed 4392 nautical miles. The fires were alight on 483 days, and she consumed 8121 tons of coal. The distance run per ton of coal was 7·93 miles, and the coal per indicated horse-power per hour 5·85 lb. Trials with expansive working were carried out in many vessels. In the discussion on a paper on "High-Speed Steam Navigation", by Robert Armstrong, read to the Institution of Civil Engineers in March, 1857, particulars were given of trials in the paddle frigates *Retribution* and *Terrible*. In the latter ship, which had Maudslays' twin-cylinder engines, seven different grades of expansion were tried, the horse-power ranging from 1850 down to 650. At 1640 indicated horse-power the coal consumption was 4·33 lb. per indicated horse-power per hour and at 650 indicated horse-power, 6·42 lb. As was pointed out at the time, the excessively high consumption was largely due to condensation in the cylinders.

Jet condensation continued to be the general practice at sea till the 'seventies, though many experiments were made with surface condensation previous to that time. The great pioneer of the marine surface condenser was Samuel Hall, of Basford, Nottinghamshire. Hall was born in 1781 and died in reduced circumstances November 21, 1863, in Morgan Street, Tredegar Square, Bow, Essex. He had been successful with inventions for gassing lace and net before turning his attention to steam engines. His patent No. 6556 of February 13, 1834, was one of the most interesting ever taken out in connection with marine engineering. It included the combination of a tubular condenser, an air pump, a circulating pump, an evaporator or still for maintaining a supply of fresh water, and a steam saver, or automatic blow-off, by which the steam escaping from the safety valve was led back into the condenser. His condenser, evaporator and steam saver are shown in Figs. 18, 19, 20.

The tubes in the condenser were fitted with screw ferrules as they are to-day. The evaporator was placed in a corner of the boiler, the steam generated in it being led to the condenser. In the steam saver the upper part *aa* was like a cartridge case and dipped into mercury contained in the annular space between *bb*. It was loaded with weights *gg*. As it was lifted by the steam it opened the valves *f* and *e*, allowing the steam from the boiler pipe *c* to flow to the condenser by pipe *d*.

On April 19, 1834, the *Mechanics' Magazine* said that

The well-tried favourite of the public the *Prince Llewellyn*, now plying twice a week betwixt the Menai Straits and Liverpool, is the first packet that has been fitted out on Mr Samuel Hall's principle for the im-provement of steam engines, consisting of a superior method of condensing the steam and using fresh in-stead of salt water, thereby creating a great saving in the boilers, and at the same time consuming one-third less of fuel.

The second vessel with Hall's condenser was the *Windermere* engined by Fawcett, Preston and Company. Tried in H.M.S. *Megaera*, and H.M.S. *Penelope* and the early transatlantic ships *Sirius* and *British Queen* and other vessels, surface condensers were shortly after-wards abandoned for some time, partly on the score of expense and partly because they became clogged with the tallow so freely used in boilers and engines at that time. It was not till after 1860 that they came into use on any considerable scale, and their reintroduction was largely due to Edward Humphrys and to John

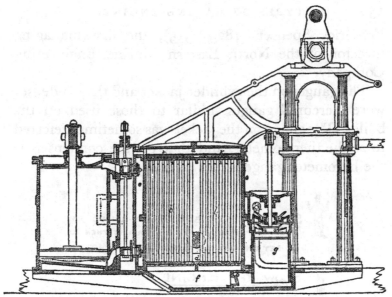

Fig. 18. Hall's surface condenser fitted to side-lever engine

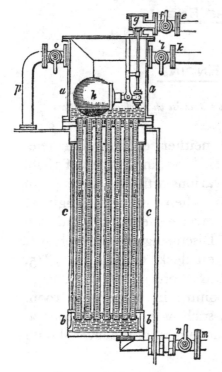

Fig. 19. Hall's evaporator

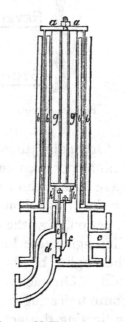

Fig. 20. Hall's steam saver

Frederick Spencer (1825–1915), the first managing director of the North Eastern Marine Engineering Company.

The gauges on the cylinder jacket and the condenser were mercurial gauges similar to those used on the boiler. The gauge on the jacket was sometimes referred to as the thermometer, and that on the condenser as the barometer gauge.

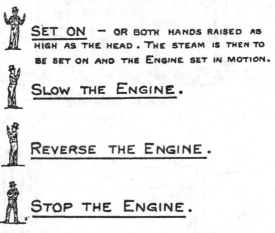

SET ON — OR BOTH HANDS RAISED AS HIGH AS THE HEAD. THE STEAM IS THEN TO BE SET ON AND THE ENGINE SET IN MOTION.

SLOW THE ENGINE.

REVERSE THE ENGINE.

STOP THE ENGINE.

Fig. 21. Sketches from Ross's *Treatise on Navigation by Steam*

Originally there were neither engine-room telegraphs nor even voice pipes. Ross, in his book of 1828, gave figures showing the various attitudes the captain on the bridge should take up when wishing the engineer to set on, slow the engine, reverse the engine, and stop the engine (see Fig. 21). Discussing the control over the engine by the officer on deck, Gordon, in 1845, said: "On each paddle box there should be a lever communicating with a pointer in the engine room indicating distinct orders, such as 1, ease; 2, stop; 3, back; 4, go on, and each movement of the lever should ring a bell. The degree of the power should be indicated by repeating the same movement." The same writer also said it would be advantageous if the officer of the watch possessed the means of stopping the engines in case of emergency. To effect this it would only be necessary to have two levers, one connected with the throttle valve and the other with the injection cocks, but the operation of reversing the engine, he thought, could not be accomplished without more complicated machinery than it would be advisable to apply. Gordon not only wished to have engine-room telegraphs, but also firing indicators.

"In the absence of any improved *mechanical* mode of feeding the furnaces", he wrote, "we shall propose a very simple method by which at least their due and equable replenishment may be ensured. Let a piece of mechanism be fixed in the stokehole which shall be worked by the engine and connected with a pointer revolving on a clock face on which circumference shall be engraved a distinguishing number corresponding to each furnace. Then when the pointer arrives at No. II, the furnace indicated must be replenished with a definite amount of fuel, which amount must be regulated by the engineer on duty...."

There are many other matters connected with early marine engineering practice which have some historical interest, but to which only the briefest reference can be made. Indicators and dynamometers for use with marine engines were dealt with in a little book by Main and Brown. Indicators had cylinders $1\frac{3}{4}$ in. diameter. The lubrication and preservation of the machinery is touched upon by most of the early writers. Otway said, "in lubricating the engines, it is found that the harder the lubricating substance, the less likely are the bearings to become overheated.  Hence, melted tallow is preferable to oil." The tallow supplied by the dockyards, for some reason, he found, contained considerable amounts of lime, and this necessitated the tallow being re-melted before being used. We know, from Commander Robinson, that engine rooms were clean, bright and well arranged, but the engineer was enjoined by the regulations "not to use, nor to permit the use of brick-dust, emery, or such other substances for polishing the bright parts of the machinery, especially those portions of it which are in the vicinity of the working parts. They are to be merely kept oiled, so that they may become bronzed by the action of the sea air." Small parts of spare gear were carried in the ships, but in the case of naval vessels the larger parts were sent to the dockyards of foreign stations on which the ships were serving, as is the practice to-day. The average daily consumption of stores for a small steam vessel in the Navy according to Otway was:

|  | Harbour | Sea |
|---|---|---|
| Sweet oil | 5 pints | $1\frac{1}{2}$ gals. |
| Tallow | $1\frac{1}{2}$ lb. | 20 lb. |
| Lamp oil | $1\frac{1}{2}$ gals. average for sea and harbour | |
| Candles | $\frac{1}{2}$ lb. | — |
| Spun yarn | — | 10 lb. per week |
| White oakum | 4 lb. | 30 ,, |
| White lead | 1 lb. | 7 ,, |
| Red lead | — | $1\frac{1}{2}$ ,, |

Other things which occupied the attention and exercised the ingenuity of many engineers, both ashore and afloat, were the design and construction of the paddle wheel and screw propellers, and the means of reefing and disconnecting the former, and the devices for feathering and lifting the latter. The experiments with these apparatus were exceedingly numerous. Some were successful and some otherwise, but from them all came some contribution to the advancement of marine engineering practice.

## CHAPTER XI

## STEAM AND SAIL FROM 1860 TO 1870

THE preceding chapters have been devoted to a general review of naval and marine engineering from the first use of steam for the propulsion of vessels down to the end of the 'fifties of last century. The birth of steam navigation, the conquest of the ocean by steam, the introduction of steam fighting ships, and the use of iron for ship-building have been among the subjects dealt with, and it has been shown how the work of the mechanical engineer has influenced sea transport and sea warfare. The materials for such a review, although abundant, have to be sought for in many and obscure directions; in continuing the survey further, the records are still more numerous, but fortunately more easily accessible. In the proceedings of certain technical societies, and in naval and engineering journals and other works, are to be found summaries of progress written by men who had a share in furthering the development of both ship-building and marine engineering. In these summaries, the outstanding landmarks in progress have been indicated again and again. But, while such is the case, the subject becomes more and more complex, and it becomes increasingly difficult to deal with every aspect. In view of this it is necessary to set closer limits to the survey, and the succeeding chapters, therefore, treat more particularly with purely engineering matters than has been previously the case. Before, however, proceeding to trace the steps which have led to the types of machinery found in ships to-day, it is of interest to take a general survey of the decade 1860–70, a period which, for many reasons, has a peculiar interest for all familiar with nautical affairs. A comparison of the fleets, both naval and mercantile, of 1860 with those of 1870, presents many striking contrasts. In 1860, in spite of the progress of steam, the greater part of the work at sea was still being done under sail. Though our mercantile marine included some 700,000 tons of steamships, the sailing tonnage was some seven times larger. Steam had invaded naval fleets to a greater extent than mercantile fleets, and this is shown by the fact that, in 1860, the Mediterranean and Channel squadrons were composed entirely of vessels fitted with steam engines. But for all that, the efficiency of the ships, both then and for some years afterwards, was judged almost entirely by smartness at sail drill, and by the handling of the ships under sail. Never in fact did British seamanship stand higher than at the time when the steam engine was first seriously threatening the existence of masts and yards. In our steam frigates, the rivalry between the crews was as keen as it was between the crews of the famous sailing clippers. Of the remarkable voyages of British and American clippers between Foochow and London, Liverpool and Melbourne, New York and San Francisco, there are many vivid accounts. Ships such as the *Lightning*, *Taeping*, *Ariel*, *Thermopylæ* and the *Cutty Sark*, are among the most famous vessels of all time. Under favourable conditions the speed of these vessels far exceeded that of any steamship. The *Lightning*, for instance, is recorded to have once sailed 420 miles in 24 hours, in 1854 the *Red Jacket* made the passage from New York to Liverpool in 13 days 1 hour, while in 1862 the *Dreadnought* crossed from Sandy Hook to Queenstown in 9 days 17 hours.

Fighting under sail had come to an end at the bombardment of Odessa, during the Crimean War, on April 22, 1854, when the *Arethusa*, 50 guns, stood in and engaged the batteries, and seven years later the last of our sailing frigates in commission returned home from the Pacific to pay off. There was, however, little diminution in the interest taken by naval officers in sail, and they were enjoined by the regulations to use steam as little as possible. One landmark in the history of sail in the Navy was set up on March 27, 1865, when for the last time a line-of-battleship, the *Edgar*, sailed out of Portsmouth harbour, while the decade 1860–70 closed with the voyage around the world of the Particular Service Squadron, all the passages of which were done under sail. Leaving England on June 19, 1869, the squadron, which consisted at first of four steam frigates and two steam corvettes, visited South America, the Cape, Australia, Japan, Honolulu and Valparaiso, returning home round Cape Horn and entering Plymouth Sound on November 15, 1870. The primary object of the cruise was the training of the officers and men in seamanship, and it mattered little whether the ships had engines or not. But no one perhaps knew better than Admiral Phipps Hornby, who was in command, how little sail would be used in future wars, and when he said goodbye to his officers and men, it was a farewell also, as his biographer says, "to wooden ships, to sails and yards, to the old Navy of Nelson's time. Henceforth came the era of steam and iron, of torpedoes and electricity; of what is called Science *versus* the keen observation which gained every advantage possible to be taken from wind and weather, and which used to be called Seamanship."

If the decade 1860–70 was a great sailing era, it was

still more an era of great changes in the character of our capital fighting ships, stimulated first by the experiences of the Russian War and secondly, by those of the Civil War in America. Our finest warship afloat in 1860 was H.M.S. *Victoria,* an unprotected wooden sailing line-of-battleship with full sail power and auxiliary steam power. She was 260 ft. long, 60 ft. broad and had a displacement of about 7000 tons. Her engines developed 4200 horse-power. Her armament consisted of 121 guns, the largest of which threw a projectile of 68 lb. weight, i.e., but a little heavier than some of the cannon balls of the days of Queen Elizabeth. The *Victoria* was launched in November 1859, and exactly ten years later, at Portsmouth in November 1869, Mr H. C. E. Childers, the First Lord of the Admiralty, clinched the first rivet in the keel of H.M.S. *Devastation,* the first of the so-called "mastless" turret ships, and the prototype of all our subsequent battleships. Designed by Sir Edward Reed, the *Devastation* was 285 ft. long and of 9300 tons displacement. Her twin-screw engines developed 6600 horse-power and gave her a speed of $13\frac{3}{4}$ knots. She was built of iron throughout, carried four 35-ton, 12-in. guns, mounted in two turrets, and was protected by 2540 tons of iron armour, the thickness of some of which was 14 inches. She carried no sails and had only a single mast for signalling and for lifting boats. The shots of the *Victoria* could pierce about 3 in. of iron, but those of the *Devastation* more than 12 in. Nothing perhaps illustrates better the momentous change in the construction and equipment of men of war during the years 1860 to 1870, due to the application of engineering science, than a comparison of the *Victoria* and the *Devastation.*

To these revolutionary changes no event contributed

more than the construction in the United States, by the eminent Swedish engineer, John Ericsson, of his earliest "monitors". When the American Civil War broke out, Ericsson was 58 years of age. His work on the locomotive, the screw propeller, the fire engine, and the caloric engine, had caused him to be known as one of the most versatile inventors of his age, and among the subjects which had engaged his attention was that of armoured fighting ships. His original drawings of a vessel with a low freeboard, fitted with a gun in a cupola were made in 1854, but it was not until 1861 that he designed the historic *Monitor*. Soon after the outbreak of the war between the North and the South, the Confederate Government decided to build an armoured vessel "which could traverse the entire coast of the United States, prevent all blockade and encounter, with a fair prospect of success, their entire navy". To achieve this object they raised the hull of the sunken U.S. steam frigate *Merrimac*, and in place of her upper works, constructed a large casement with sloping sides of timber protected with iron bars. In this casement were mounted two 7-in., two 6-in., and six 9-in. guns. So equipped, she steamed into Hampton Roads and, on March 8, 1862, successfully engaged the Federal fleet. While the Confederates had been busy altering the *Merrimac*, the Federal Government, however, had not been idle, and in October 1861, had accepted the offer of Ericsson to build "an impregnable battery" in 100 days. Her keel was laid down on October 25, and this vessel, called by Ericsson himself the *Monitor*, was completed by the middle of February 1862. On March 9, the famous duel took place between her and the *Merrimac*, a fight which went far to save the North, and which, said one writer, "created the profoundest sensation in the court of every maritime nation". The *Monitor* was built of iron and was 175 ft. long, and 41 ft. wide. Her displacement was about 1200 tons, her draught 11 ft., and her freeboard about 2 ft. Her armament consisted of two 11-in. Dahlgren guns mounted in a single turret, the sides of which were built up of eight thicknesses of 1-in. iron plates. She was driven by steam and steam turned her turret. She was an assemblage of many inventions rather than a single invention, and in her construction the entire resources of engineering knowledge of the time were brought to bear on the solution of a practical problem. "The *Monitor*", said Scott Russell, "is a creation altogether original, peculiarly American; admirably adapted to the special purpose which gave it birth. Like most American inventions, use has been allowed to dictate terms of construction, and purpose, not prejudice, has been allowed to rule invention." Imperfect in many respects, as the *Monitor* proved to be, there was not a single navy which did not utilise some of the ideas first put into practice in her, in 1862, by Ericsson.

While the war in America hastened the fundamental change in the design of fighting ships which took place in the decade 1860–70, it contributed also to the advancement of marine engineering by the demand for fast ships capable of running the blockade and for ships for the protection of commerce or the destruction of hostile vessels. One British firm which came into prominence in this connection during the 'sixties was that of J. and W. Dudgeon, of Millwall. Both John Dudgeon (1816–81), and William Dudgeon (1818–75) began life as blacksmiths in Scotland, but having gained experience with some of the leading engineering firms, in 1859 they founded an engineering shop at Millwall and a year or two later started shipbuilding at Cubitt Town. One of the first vessels they engined was the *Thunder*, of 1062 tons, which attained a speed of 14 knots. In 1862, they built and engined the *Flora* of 450 tons and 120 nominal horse-power, with independently driven twin screws. So successful did this vessel prove that she was diverted from her original purpose to become a blockade-runner. Twin screws had been tried by Stevens as early as 1804, and had also been patented, but the Dudgeons were the first to build any considerable number of twin-screw vessels and to demonstrate fully the value of the plan. In a paper contributed by the brothers to the Institution of Naval Architects, in 1865, they said: "Since March 1862, we have completed twenty twin-screw steamers, of varying tonnage and power, up to 1500 tons and 350 horse-power. First, we built eight vessels of 425 tons and 120 horse-power, all destined for blockade running. They have all been very successful in that hazardous trade, some running as many as a dozen times, the average number of runs being five." The speed of the vessels was 14 knots and they possessed superior manœuvring power. The Dudgeons added further to their reputation by the construction of the *Far East* and the *Ruahine*, the latter of which had twin screws driven by compound engines.

The necessity of countering the activities of these blockade runners and other vessels led the Federal Government, in 1863, to lay down several ships designed to have a speed of 17 knots. They were built of wood, 341 ft. long, and with a displacement of 4000 tons. They were to carry sixteen 10-in. and 11-in. smooth-bore cast-iron guns on the broadside, and a 60-pounder rifled gun in the bows. Owing to their

great length, the hulls were strengthened with longitudinal and diagonal stringers of iron plate. At the
time of their construction, Benjamin Franklin Isherwood (1822–1915) was the Chief Engineer of the United
States Navy. One of the first naval engineers to apply
scientific methods to the trials of engines, he had published his *Engineering Precedents* in 1859, and had carried
out many trials with marine machinery. In his own
country he has been spoken of as "the greatest marine
engineer the United States has produced". He was
responsible for the design of the machinery fitted in the
fast vessels referred to, and although some of the vessels
did not come up to expectations, the *Wampanoag* gained
for herself a place in history as the fastest ship of the
time, attaining a speed of 16·7 knots. Her machinery
consisted of a two-cylinder simple-expansion engine,
with cylinders 100 in. diameter and 4-ft. stroke, driving
the propeller through wooden-toothed gearing. On
trial, the engine speed was 31·06 revolutions per minute
and the propeller speed 63·67 revolutions per minute;
the power developed was 4048 indicated horse-power.
The *Wampanoag* and her sisters had a very short life, and
were condemned within a year of their trials, among
the reasons given for this action being "excessive
amount of fuel burned" and "the grate surface exceeded the area of the midship section of the ship".
Isherwood had to face a great deal of criticism at the
time, but the vessels he engined were considered to be
the finest of their kind in existence at the end of the war,
and they introduced the era of the fast cruisers. The
earliest British vessel of this class was the *Inconstant*,
launched at Pembroke, on November 1, 1868. Unlike
the American ships, she was built of iron and sheathed
with wood and copper. She was 337 ft. long, of 5780

tons displacement, and had simple-expansion engines by Penn, developing 7360 indicated horse-power. She attained a speed of 16½ knots on trial, and her steaming capacity was well shown on the occasion of the disaster to H.M.S. *Captain*, in 1870, when the *Inconstant* was sent home with the news, maintaining on passage a speed of 15¾ knots.

The speeds attained by the *Wampanoag* and the *Inconstant* were considerably higher than those of the large ocean-going mail steamers of the 'sixties, among the most notable of which were the *Scotia* (1862) and *Russia* (1867) of the Cunard Line, the *City of Paris* (1866), and the *City of Brussels* (1869) of the Inman Line, and the *Pereire* (1866) of the Compagnie Générale Transatlantique. All of these vessels held the record for the passage of the Atlantic at one time or another, and the result of the work of the decade was to increase the average sea speed from 13½ knots to 14½ knots, and to reduce the time of crossing from about nine days to a little under eight days. The *Scotia* was a paddle-wheel vessel, while the others were single-screw ships, but all of them were fitted with simple-expansion engines, which were the results of the slow development of half a century. Though the Peninsular and Oriental Steam Navigation Company and the Pacific Steam Navigation Company, during the 'sixties, installed compound engines in many of their vessels, such engines were still in the experimental stage, and were used only to a small extent in the Navy or in the Atlantic liners. The history of the introduction of compound working, probably the most important improvement in marine machinery made during the nineteenth century, and the use of higher steam pressures, is dealt with in the next chapter.

Among the experiments on the propulsion of ships made during the 'sixties, which by some engineers were considered to be of much promise, were those with jet propulsion in S.S. *Nautilus* and H.M.S. *Waterwitch*. The idea of drawing water in at the bow of a ship or through the bottom and discharging it in a stream at the stern is very old, and goes back to 1661. It was one of the methods tried by James Rumsey, the American pioneer, and prior to 1859 no fewer than 59 patents had been taken out in connection with it. The chief advocates of the system were J. and M. W. Ruthven, of Edinburgh, whose patent was taken out in 1839. Their system was tried in the fishing vessel *Enterprise* in 1854, and again in 1866 in the *Nautilus* and *Waterwitch*. Many years later the Admiralty had a torpedo boat constructed for hydraulic propulsion, and much information regarding the subject is contained in a paper by S. W. Barnaby, published in the *Proceedings of the Institution of Civil Engineers*, 1883–4. The *Nautilus* was 115 ft. long, and with machinery of 127 indicated horse-power, attained a speed of 8·32 knots. The *Waterwitch* was considerably larger, being a gunboat 162 ft. long, of 32 ft. beam and 1161 tons displacement. The machinery constructed by Dudgeons to the Ruthvens' design (see Fig. 22) consisted of a three-cylinder engine with a vertical crankshaft, at the lower end of which was the impeller of a large centrifugal pump. The impeller weighed 8 tons and revolved in a chamber 19 ft. in diameter. The machinery was placed in the middle of the hull, and water was drawn through a large number of openings in the bottom and discharged through two nozzles 24 in. by 18 in., one on either side of the vessel, fitted just above the water line. By sluice valves the water could be directed ahead or astern. With the engine developing 777 indicated horse-

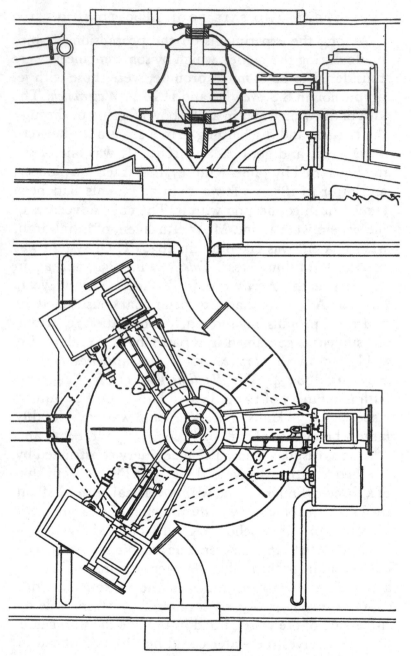

Fig. 22. Longitudinal section and plan of machinery of
H.M.S. *Waterwitch*

power, the speed of the vessel on trial was 9·3 knots, but in the open sea only 5 to 6 knots. Both this and all similar experiments demonstrated the inefficiency of the method, but in spite of this, there are certain cases where it can be used to advantage.

While the foregoing paragraphs recall some of the principal achievements and experiments of the decade 1860–70, there is another aspect of the activities of the time which deserves attention. This was the increasing attention shown by marine engineers to theory, and the fuller realisation of the interdependence of abstract science and engineering progress. The growth of this is worth tracing. During the first half of the nineteenth century, progress in engineering was mainly due to the work of men of unusual ability risen from the ranks of the mechanics and millwrights. Stephenson, Fairbairn, Telford, the Napiers, Laird, the Maudslays, Whitworth, Field, Penn, the Hawthorns and many others, all came from the bench or the engine house. Whatever scientific knowledge they possessed was gained by private study. In his presidential address to the Institution of Civil Engineers in 1866, Sir John Fowler remarked "We of the passing generation have had to acquire our professional knowledge as best we could, often not till it was wanted for immediate use, generally in haste and precariously, and merely to fulfil the purpose of the hour." When the locomotive and steam boat came into existence there was no teaching of engineering science, technical education was in its infancy, and the application of scientific research to practical problems was scarcely thought of. Much mathematical talent, it is true, had been applied to certain subjects, especially by the French, and during the eighteenth century the French had established government schools of civil engineering, mining and hydrography. The French text-books too, were regarded as excelling those in any other language and even in a matter of such national importance as the design of warships, our constructors had to seek in the works of foreigners for discussions of the scientific principles underlying their work.

Technical education in Great Britain may be said to have had its birth about a hundred and fifty years ago when Professor John Anderson lectured to the artisans of Glasgow. After his death, the Andersonian Institution, the forerunner of the Glasgow Royal Technical College, was founded and this in turn led to the establishment in many parts of the country of the mechanics' institutions of a century ago. Birkbeck College is one of the survivors of these institutions. In 1811 the Admiralty opened the first British School of Naval Architecture at Portsmouth. This was followed in 1837 by the promulgation of orders for the training of engineers for the Navy and in 1843 by the opening of the dockyard schools for apprentices. The earlier school of naval architecture having been closed in 1832, a second one was opened in 1848 but this had only a short career.

By this time engineering professorships had been established, first at University College, London, then at Dublin and Glasgow, and the publication of investigations affecting the work of the shipbuilder and engineer had become common. The significance of some of these inquiries was not realised till long afterwards. Various researches into the strength of materials were made by Barlow, Tredgold, Hodgkinson, Tate and others, but foremost among the researches which subsequently influenced the progress of marine engineering were those on thermodynamics by Carnot and Joule. No single individual perhaps had more influence on his contemporaries than William John Macquorn Rankine (1820–72), professor of engineering in Glasgow University, declared by Unwin to be "the first really powerful thinker in this country to bring the highest mathematical resources to bear on engineering practice".

In no department of practical work was a knowledge of scientific principles found of greater value than in the construction of ships, and it was this which led to the founding in 1860 of the Institution of Naval Architects. To further their plans for improving technical education, the promoters of this society prevailed upon the Admiralty in 1864 to open the Royal School of Naval Architecture and Marine Engineering at South Kensington. A good many shipwright apprentices had passed through the earlier schools mentioned above, but the school at South Kensington was the first in this country for training marine engineers.

The school was in existence from 1864 to 1873, when its work was transferred to the Royal Naval College, Greenwich. During those nine years, among its engineer students, all of whom had been trained in the dockyards for service in the Fleet, were Bedbrook, Durston, Sennett, Hearson, Morcom, Butler, Chilcott, Mayston, Corner, Waghorn, Seaton and Milton who all rose to high positions in the marine engineering world. The course of training was severe, the standard of examination high. Among the text-books in use were the mathematical works of Boole, Besant, Todhunter and Routh, and Rankine's *Steam Engine and other Prime Movers*, the first engineering work to contain an explanation of the new science of thermodynamics which was to play so important a part in the subsequent development of the marine steam engine.

## CHAPTER XII

# THE MARINE COMPOUND ENGINE

IN the last chapter a brief reference was made to the introduction into ships of the compound steam-engine in which the steam is expanded in two stages. This was a step in the progress of marine propulsion which had most important results. It not only led to a great reduction in the amount of coal consumed but, necessarily accompanied by the use of steam at higher pressures, paved the way for increased speeds and a reduction in the weight of machinery and the space it occupied. With simple-expansion engines, working at about 20 lb. steam pressure and fitted with jet condensers, the coal consumption of many vessels was over 4 lb. per indicated horse-power per hour. By the use of compound engines working with steam at about 60 lb. pressure, and fitted with surface condensers, this was reduced to 2 lb. or even less. The effect of this reduction enabled the steam ship to become a successful rival of the sailing ship on the longest voyages, while it gave scope to the naval designer to increase both the offensive and defensive power of warships. Like all innovations in marine propulsion, the introduction of the compound engine gave rise to much discussion, and the controversy was carried on long after the value of the new type of engine had been definitely established. The first successful application of the compound engine to a sea-going ship was made in 1854, yet 20 years after that date comparative trials of vessels with either simple-expansion engines or compound engines were still being made by the Admiralty, by the United States Coast Survey and by the Allan Steamship Company. At first, the problem of determining the relative efficiency of the two types was seldom attacked on scientific lines, and the issue was frequently confused by considering simultaneously various other subjects, such as steam pressures, superheating, steam jacketing, and surface condensing. Generally speaking, the compound engine was regarded as in the experimental stage in the 'sixties, it was almost universally adopted in the 'seventies, and it remained the most popular type of engine for ship propulsion till superseded by the triple-expansion engine in the 'eighties.

The doubts which long existed in the minds of some engineers as to the superior merits of the compound steam engine, found expression on many occasions. When, after the *Captain* disaster of 1870, a Committee on Designs of Ships of War was appointed, among the questions referred to it was that of the machinery of warships. The few naval vessels which up till that time had been fitted with compound engines had not been very successful, but, after hearing the views of various eminent engineers, the Committee reported that the weight of evidence in favour of the large economy of fuel effected by the compound engine was overwhelming and conclusive. The Report of the Committee was issued in July 1872, and immediately it met with criticism. Discussing the question of machinery, a writer in *Naval Science* for October, 1872, said:

As an engine peculiarly adapted for high-pressure steam, we regard the general introduction of the compound engine in the merchant and mail services and in special service ships attached to our Navy as only a question of time; but that any advantage would be gained by its extensive adoption for our ships of war as strongly recommended by the Committee on Design, appears to be very doubtful.

That the writer in *Naval Science* was not alone in the views he held can be seen by glancing through the discussions which took place after the reading of papers on compound engines before the technical societies. Two of the most valuable of such papers were Sir Frederick Bramwell's paper entitled "On the Progress Effected in Economy of Fuel in Steam Navigation", read before the Institution of Mechanical Engineers, at Liverpool, July 30, 1872, and Alfred Holt's paper on "Progress of Steam Shipping during the Last Quarter of a Century", read before the Institution of Civil Engineers, in November 1877. The discussion of Holt's paper extended over four evenings and the report occupies 130 pages in the *Proceedings*. Holt had been an engineer before he became a shipowner, and was himself a pioneer in the use of the compound engine at sea, yet after ten years' experience with it he said, "it is a matter of reasonable speculation, whether the compound may not yet be abandoned, and a return made to the single-cylinder engine, modified in details to suit high-pressure steam". By this term, Holt meant steam at about 60 lb. pressure, which had been declared by Macfarlane Gray in 1872 to be as high a pressure as could be advantageously used at sea. In reading the discussion, it is seldom found that the theoretical principles underlying the economy of compound engines were understood. Mr (now Sir) Fortescue Flannery, however, in dealing with Holt's paper, pointed out the importance of the reduction of the range of temperature occurring in any one cylinder, while Richard Sennett, then fast rising to distinction, said that the advantages of compound working were: (1) reduction of the maximum strains in the framing, shafting and bearings and consequent reduction of weight and cost;

(2) increased regularity of turning moment and consequent increased efficiency of the propeller in the water; (3) more economical use of the steam in the cylinders, and consequent increase of power from a given expenditure of heat.

The plan of expanding steam successively in two cylinders, a smaller high-pressure one, and a larger low-pressure one, goes back to the days of Watt, and was first used in beam engines long before the theory of heat engines was understood. The younger Jonathan Hornblower patented such an arrangement in 1781, and the following year erected an engine at Radstock near Bath having cylinders 19 in. and 24 in. diameter. He was then stopped from proceeding further, as he was infringing the separate condenser patent of Watt. The idea lay fallow till 1803, when Arthur Woolf (1766–1837) followed up the work of Hornblower and built compound engines both in England and in France. In the latter country, the compound engine was actually known as a *machine de Woolf*. Another step in compound working was taken in 1845, by William McNaught (1813–81) who, confronted by the problem of increasing the power in cotton factories to meet extensions without scrapping the existing beam engines, added a high-pressure cylinder near the crank end of the beam. Such an addition to an engine was called "McNaughting", a term which later on was superseded by "compounding." A distinctive feature of Woolf engines, whether they had beams or not, was that the two pistons worked together in the same direction, and with this arrangement the steam exhausted directly from the high-pressure cylinder into the low-pressure cylinder, no intermediate steam receiver being required. From what has been said, it will be seen that the compound engine first made its

appearance on land. In the 'thirties and 'forties, it began to be fitted to small vessels, but it was not till John
Elder began his work in the 'fifties that the compound
engine came to be used for sea-going ships. His work,
indeed, proved revolutionary.

Elder was born at Glasgow, on March 8, 1824, being
the third son of David Elder (1785–1866), who for
many years was manager for Robert Napier. Educated
at Glasgow High School, young Elder served an apprenticeship to Napier, and about the age of 24 became his
chief draughtsman. In 1852 he became a partner with
Charles Randolph and thus founded a firm at Govan
first known as Randolph, Elder and Company, then as
John Elder and Company, and lastly as the Fairfield
Shipbuilding and Engineering Company. He died
on September 17, 1869, at the early age of 45.
He had, however, done notable work in connection
with the construction of ships and engines and is
generally regarded as the chief pioneer of the marine
compound engine. When in 1888 his statue in Govan
Park was unveiled, it was said "by his many
inventions, particularly in connection with the compound engine, he effected a revolution in engineering
second only to that accomplished by James Watt,
and in great measure originated the developments
in steam propulsion which have created modern
commerce".

In January 1853, Elder and Randolph took out jointly
a patent "for an arrangement of compound engines
adapted for the driving of the screw propeller. The engines are vertical, direct-acting and geared. The pistons
of the high- and low-pressure cylinders move in contrary directions, and drive diametrically opposite cranks
with a view to the diminution of strain and friction".

PLATE VII

BENJAMIN FRANKLIN ISHERWOOD
(1822–1915)
Courtesy of American Society of
Mechanical Engineers

CHARLES SELLS
(1820–1900)
Photo by Maull and Fox

JOHN ELDER
(1824–69)

CHARLES-BENJAMIN NORMAND
(1830–88)

This patent was followed by others relating to the same subject, and in 1862 Elder secured a patent covering three-cylinder triple-expansion, and four-cylinder quadruple-expansion engines "for steam of very high original pressure". The first set of engines constructed under the original patent was fitted in the *Brandon*, which on trial was found to have a coal consumption of 3¼ lb. per indicated horse-power per hour. Two years later, the Pacific Steam Navigation Company, whose vessels were employed on the west coast of South America where coal was expensive, placed an order with the firm to fit compound engines in the new paddle-wheel vessels *Inca* and *Valparaiso*. The success of these vessels led to the *Lima*, *Bogota* and *Callao*, of the same company, being sent home to be re-engined with compound engines. Of the work of these ships, Elder gave accounts to the British Association in 1858 and 1859, and it was stated that the *Bogota's* coal consumption had been reduced more than 50 per cent. by the alteration. The steam pressure used was only about 25 to 30 lb., and some of the improvement in performance was due to the attention paid to steam jacketing. The success due to compounding, however, was undeniable, and by 1866 Randolph and Elder had built 48 sets of compound engines, 18 for paddle wheels and 30 for screw engines, most of the engines being for ships employed in the Pacific.

The second steamship company to adopt the compound engine was the Peninsular and Oriental Steam Navigation Company, to which economy of fuel was just as important as it was to the company already mentioned. Many of the vessels of the P. and O. Company were built on the Thames, and the task of fitting them with compound engines fell to Edward Humphrys, of

Humphrys, Tennant and Company. Elder had used cylinders placed side by side, but Humphrys introduced the use of cylinders placed tandem-wise. Beginning with the *Mooltan* (1861), he supplied compound engines to the *Carnatic* (1862), the *Poonah* and *Golconda* (1863), and the *Baroda* (1864). The fine model of the engines of the *Carnatic*, in the Science Museum, shows the type introduced by Humphrys. The engines were vertical and inverted with two high-pressure cylinders placed above two low-pressure cylinders, driving a two-throw crank-shaft. The stroke was 3 ft. With steam at 26 lb. pressure, the indicated horse-power was 2442, and the coal consumption under 2 lb. per indicated horse-power per hour. By 1866 the P. and O. Co. had ten vessels running with compound engines. Another company which achieved success with ships fitted with compound engines was the Ocean Steam Ship Company, or Blue Funnel Line, founded by Alfred Holt. His three ships, *Agamemnon*, *Ajax* and *Achilles*, of 1866, had vertical tandem engines, with one high-pressure cylinder and one low-pressure cylinder, driving a single crank. Holt adopted this type to save weight and space. The constructors of the engines were Scott, of Greenock, and R. Stephenson and Company. Holt's vessels traded to the Far East, and they were able to make the run of 8500 miles from England to Mauritius without recoaling. With the opening of the Suez Canal in 1869, the difficulty of direct communication with the Far East and Australia was, of course, diminished considerably but, in view of the cost of coal abroad, the need for economy still remained. The effect on sea transport of the shortening of the passage and the improvement in the performances of steam ships fitted with compound engines was quickly seen, and it was the combination of these improvements which sounded the knell of the famous China sailing clippers.

While the compound engine thus came into use during the 'sixties for ships on certain routes, it made but slow progress in Atlantic liners or in the ships of the Royal Navy. The *City of Rio de Janeiro*, engined by Elder, started running to Brazil in 1868, and the following year the National Steamship Company adopted the compound engine for the *Italy* and *Holland* running to New York. The same year the Cunard Company used the compound engine in the *Batavia* and *Parthia*, and from 1870 onwards, practically all transatlantic vessels were fitted with compound engines. The types used were those with the cylinders placed side by side, those with cylinders placed tandem-wise, and those with one high-pressure and two low-pressure cylinders side by side, which latter appears to have first been used in France. All the engines were vertical. The greatest development of the compound engine was seen in the single-screw vessels *Etruria* and *Umbria* of the Cunard Company, which had engines of no less than 14,500 indicated horse-power, the steam pressure being 110 lb. per square inch. These vessels were built in 1884, and they were the last of a group of Atlantic liners by which the passage was gradually reduced from eight days to less than six days. Of six of these vessels, a few particulars are given in Table X on p. 182. The engines of the *Britannic* and of the *City of Rome* had the high-pressure and low-pressure cylinders tandem, but in the engines of the others the cylinders were side by side.

The decision to try the compound engine in the Navy was taken as early as 1860, when the three wooden sailing frigates *Arethusa*, *Octavia* and *Constance*, were being converted to screw ships. Built in the 'forties, the ships were 252 ft. long and about 3100 tons displacement. For the *Arethusa*, Penns supplied simple-expansion trunk engines; for the *Octavia* Maudslays supplied simple-expansion return connecting-rod engines, while for the *Constance* Randolph and Elder constructed an engine with two high-pressure and four low-pressure cylinders driving a three-throw crankshaft, the cylinders being

## Table X

| Date | Ship | Length ft. | Gross tonnage | Boiler pressure lb. | Cylinders No. diam. | Stroke ft. | I.H.P. | Speed knots |
|------|------|-----------|---------------|---------------------|---------------------|-----------|--------|-------------|
| 1874 | *Britannic* | 455 | 5004 | 70 | 2–48 2–83 | 5 | 4,971 | 13 |
| 1879 | *Arizona* | 450 | 5164 | 90 | 1–62 2–90 | 5½ | 6,357 | 17·3 |
| 1881 | *City of Rome* | 560 | 8141 | 90 | 3–43 3–86 | 6 | 11,890 | 18·2 |
| 1881 | *Servia* | 515 | 7390 | 90 | 1–72 2–100 | 6½ | 10,300 | 17·8 |
| 1881 | *Alaska* | 500 | 6932 | 100 | 1–68 2–100 | 6 | 11,000 | 17¾ |
| 1884 | *Umbria* | 520 | 8120 | 110 | 1–71 2–105 | 6 | 14,500 | 19 |

disposed in V-form. There was delay in completing the vessels, but ultimately they met in Plymouth Sound in 1865, for a trial race to Madeira. A start was made on September 30, and the race practically came to an end on the afternoon of October 6, when the ships were running short of coal. At that time the *Constance* was about 30 miles from Funchal, the *Octavia* 160 miles, and the *Arethusa* 200 miles. The report of the trial showed that the *Arethusa's* coal consumption was 3·64 lb. per indicated horse-power per hour, the *Octavia's* 3·17 lb. and that of the *Constance* 2·51 lb. Particulars of the engines of the *Constance* are given in Vol. xxxiii of the *Transactions of the Institution of Naval Architects*, and ex-

PLATE VIII

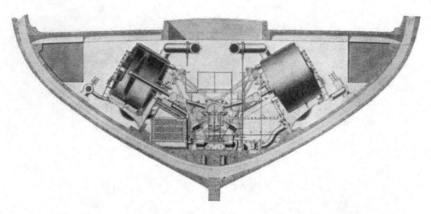

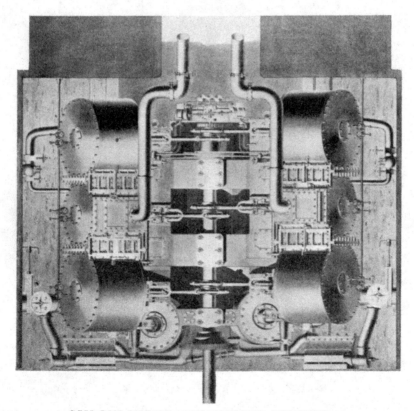

SIX-CYLINDER COMPOUND ENGINE
H.M.S. *CONSTANCE*, 1863
Courtesy of The Fairfield Shipbuilding and Engineering Co., Ltd.

ceptionally well-executed drawings of them are preserved at Fairfield. Photographs of two of these drawings are given in Pl. VIII. The engines showed that
they were undoubtedly more economical in fuel than
the engines of the sister ships, but that was their only
merit. They were complicated, difficult to handle, un-
reliable on service, and of such a character that it was
almost beyond the capacity of the engine room staff to
keep them in any way efficient.

The second ship in the Royal Navy to have compound engines was the *Pallas*, an armoured wooden
corvette of 3790 tons displacement. For her Humphrys constructed a horizontal two-crank engine with
tandem cylinders, the low-pressure cylinders being
placed nearest the crank, and being fitted with trunk
pistons in order to obtain a sufficiently long stroke. The
high-pressure cylinders were 51 in. diameter, the low-
pressure $99\frac{1}{2}$ in. diameter, the common stroke 3 ft. 3 in.,
and the boiler pressure 32 lb. per square inch. At 78
revolutions per minute, the engines developed 3200
horse-power, giving the ship a speed of 13 knots. Engines
of a similar design were also fitted in the new Indian
troopships *Serapis* and *Crocodile*, but soon gave trouble.
In one of the vessels, after running 31,000 miles, the
high-pressure cylinder was found to be worn down
$\frac{3}{8}$ in. and the cylinder had to be rebored. Trouble was
also experienced with the gudgeon bearings in the
trunks and with the wrought-iron casings forming the
cylinder jackets. Although the coal consumption was
more or less satisfactory, the cost of upkeep was so great
that the compound engines were very soon replaced by
simple-expansion engines.

With the results of these experiments before them, it
is not a matter for surprise that naval engineers showed

little enthusiasm for the new type of engine. A further trial was made in the late 'sixties with the five single-screw wooden corvettes *Briton*, *Spartan*, *Sirius*, *Tenedos* and *Thetis*. The engines constructed were all of 350 nominal horse-power, and all were horizontal, but they all differed in design. The engines of the *Sirius* were the first compound engines supplied to the Navy by Maudslays. They had two high-pressure cylinders and two low-pressure cylinders arranged tandem with the high-pressure cylinders nearest the cranks, and recessed into the front of the low-pressure cylinders, a plan adopted successfully later on in the *Iris* and *Mercury* (see Fig. 24). With steam at 55 lb. pressure, at 96 revolutions per minute the indicated horse-power was 2325. For the *Tenedos*, Elders supplied an engine with one high-pressure cylinder and one low-pressure cylinder, placed side by side. Externally the cylinders appeared to be of the same diameter, as all around the high-pressure cylinder there was a large receiver, into which the high-pressure cylinder exhausted. The steam pressure was 60 lb., and, at 99 revolutions per minute, the indicated horse-power was 2018. Of the five vessels, the *Briton* proved the most economical. Her engines, by Rennies, had one high-pressure and one low-pressure cylinder, in between which was a reheater designed by E. A. Cowper (1820–93). On trial, when developing 2149 horse-power, the coal consumption was 1·98 lb. per indicated horse-power per hour, and when developing 660 horse-power, only 1·30 lb. Of the engines of the *Spartan*, designed by A. E. Allen, it is perhaps sufficient to say that a writer in the *Annual* of the Royal School of Naval Architecture and Marine Engineering for 1872 said "that they bear the palm for inefficiency among the compound type in the Navy".

The vessels just referred to had all been completed when the Committee on Designs recommended the fitting of compound engines in all future ships. It was, however, seen that further investigations were desirable, and during the course of the next year or two, trials were carried out in some gunboats, first in the *Swinger* and *Goshawk*, and then in the *Mallard*, *Sheldrake* and *Moorhen*, some of which had simple-expansion engines and some compound engines. Accounts of the trials were placed before the Institution of Naval Architects by Sennett, and they are summarised in Seaton's *Manual of Marine Engineering*. In America also, Charles Emery (1838–98) made exhaustive trials with vessels of the United States Coast Survey. In all these trials, more attention was paid to the measurement of feed water than had hitherto been the practice, thus eliminating differences that arose through variations in the condition of the boilers, the quality of the coal and the efficiency of the stoking.

By this time, the undesirable features found in most of the early compound engines had been eliminated, and with improvements in design, difficulties disappeared, and engines became more reliable. The three-cylinder horizontal engine was successfully used in the single-screw vessels *Rover* (1874), *Boadicea* (1875), and *Bacchante* (1876), the two latter having engines of 5250 horse-power, and attaining a speed of 15 knots. All previous records were, however, broken with the machinery of the fast despatch twin-screw ships H.M.SS. *Iris* (1878), and *Mercury* (1879). These vessels attained the then exceptional speed of $18\frac{1}{2}$ knots, the ratio of horse-power to displacement being over two, while the weight of machinery was only about 280 lb. per indicated horse-power. The vessels were 300 ft.

long, 46 ft. wide, and 3750 tons displacement. They were the first vessels in the Navy constructed of mild steel and steel was used for the boiler shells, the propeller shafts, and the cylinder liners. The engines, by Maudslays, resembled generally those of the *Thetis* and were designed by Charles Sells (1820–1900), head of Maudslays' drawing office for half-a-century. Each engine had two high-pressure cylinders, 41 in. diameter, and two low-pressure cylinders 75 in. diameter, with a stroke of 3 ft. The boiler pressure was 65 lb. per square inch, and at 96 revolutions per minute the engines developed about 7700 horse-power. Before the speed of $18\frac{1}{2}$ knots was attained, several sets of propellers were tried. These were of gun-metal, the blades were polished and the securing nuts were covered by cones. The ships each carried ten 64-pounder rifled guns, but were unarmoured. Since many officers considered that to go into action with steam at 65 lb. pressure in the boilers might lead to disaster, arrangements were fitted so that all eight cylinders could be worked on the simple expansion principle, the steam pressure being lowered to about 4 lb. per square inch above atmospheric pressure. Drawings of one of the boilers of the *Iris* are given in Fig. 23, while Fig. 24 is a plan of the engine room.

The retention of horizontal engines in naval vessels for so many years had been dictated by the necessity of placing the machinery below the water-line for protection from gun fire. With, however, the general adoption of twin-screws, driven by two sets of engines, and the extended use of side armour, it was found possible to fit vertical engines in the larger vessels, and thus the design of the later compound engines in the Navy followed more closely that of the engines of the Atlantic liners, many battleships being fitted with the three-cylinder type.

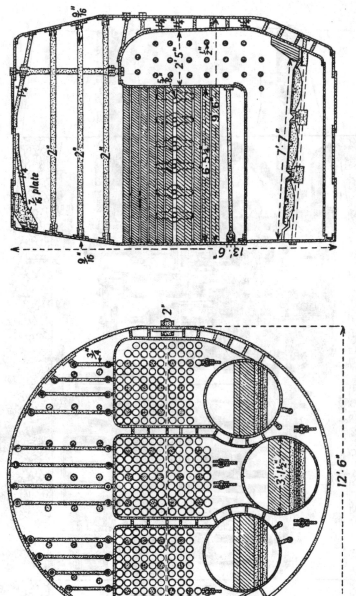

Fig. 23. Oval boiler of H.M.S. *Iris*, 1878.   From a contemporary coloured sketch

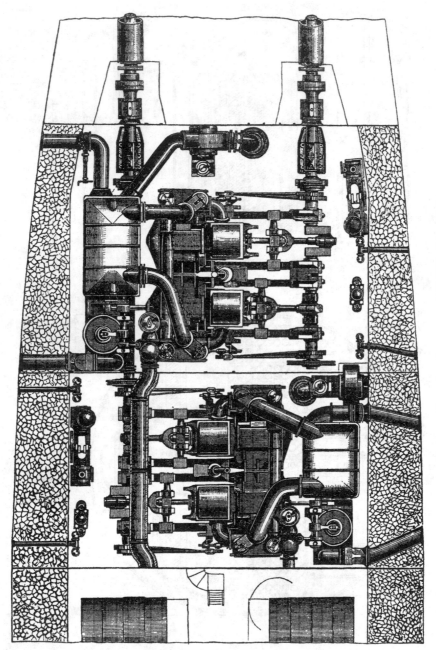

Fig. 24. Plan of engine room of H.M.S. *Iris*, 1878.

## CHAPTER XIII

# BOILER PRACTICE AND PROGRESS

URING the period of thirty years from 1854 to 1884, which practically covers the introduction and general adoption of the compound engine for ships, the working steam pressure had gradually been raised from about 25 lb. per square inch to 110 lb., or even more. This increase in pressure had brought about a radical change in the design of marine boilers, and just as the rectangular flue boiler had previously given place to the rectangular multitubular boiler, often spoken of as the "box" boiler, so this was superseded by the cylindrical multitubular boiler. Of both the box boiler and the cylindrical boiler there were two distinct types, known as the high boiler and the low boiler. In the former type the tubes were in nests above the furnaces; in the latter they were on the same level with them. Other appellations of the two types, were return-tube boilers and through-tube boilers. In all boilers the tubes were horizontal or nearly so. A variant of the box boiler was the Cochran or Martin boiler which had nests of vertical tubes through which the water circulated instead of the gases, and this was sometimes referred to as a water-tube boiler, although it had few features in common with the ordinary water-tube boiler. During the same period of 1854–84, however, many experiments were made afloat with water-tube boilers, and, with the introduction of torpedo craft, the locomotive boiler also was made use of at sea. The primary object of this innovation was to obtain the maximum amount of steam with the most compact boiler, and to assist in doing this, boilers were for the first time worked regularly under forced draught, an innovation which in itself marks a definite stage in the progress of marine engineering.

Although during the 'sixties steel, made by the Bessemer and other processes, had been used successfully by makers of land and locomotive boilers, in only a very few instances had steel been used for marine boilers which, until the introduction of Siemens steel (see p. 195), were mainly made of wrought iron. The most suitable iron came from Staffordshire and Yorkshire, and the names of Bowling, Lowmoor and Farnley and such terms as "best best" iron and "best best best" plates, indicating the number of times the iron had been reheated and rolled, were familiar in all boiler shops. The Admiralty required that iron for naval boilers should have an ultimate strength of 22 tons per square inch with the grain, and 18 tons across the grain. But iron was far from satisfactory as a material, for its very formation by the process of aggregation led to risks of lamination and defects difficult to discover; lamination in plates exposed to the heat of the fire led to the formation of blisters. The strength of a box boiler depended very much on the stays of which there was a large number holding the sides, top and bottom together. The furnaces, which were approximately square in section, also had to be thoroughly stayed. So numerous indeed were the stays that it was with difficulty men could move about inside the boilers to clean the interiors. The ordinary box boiler as found in the Navy had four or five furnaces, but in merchant vessels larger boilers were sometimes fitted, and the four boilers of the Cunard liner *Russia*, engined in 1866, by J. and G. Thomson, each had seven furnaces. Tubes of boilers in the mercantile marine were usually of iron, but those in naval vessels were of brass. Sketches of two box tubular boilers with uptakes in the boilers are shown in Fig. 25.

When boilers were in use continously and were fed with sea-water, by paying strict attention to the density

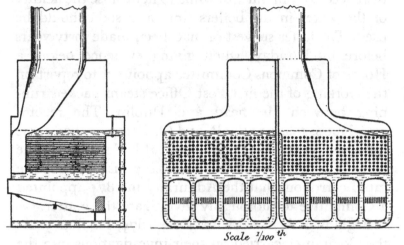

Scale ¹/₁₀₀ ᵗʰ

Fig. 25. Rectangular low-pressure tubular boilers

there was not more corrosion than might have been expected under the circumstances, and boilers sometimes lasted as long as twelve or fourteen years. With the introduction of the surface condenser and the use of fresh water there was a marked increase in the amount of corrosion, and this was especially the case in the Navy. The conditions, it is true, were often such as to lead to rapid wasting, for boilers were frequently in store a long time before being put into service; in ships in commission they were worked intermittently; they were sometimes used as ballast tanks or as fresh-water tanks, and they were frequently left open to the atmosphere. In 1862, the Peninsular and Oriental Steam

Navigation Company, owing to the rapid decay in the boilers of some of their ships, sought the assistance of the eminent chemists, G. H. Ogston and Professor A. W. Hofmann. As a result of their investigations it was recommended that for the internal lubrication of engines, tallow and vegetable and animal oils should be displaced by mineral oils, while to neutralise the acidity of the water in the boilers lime and soda should be used. The latter suggestion had been made forty years before by Faraday, when giving evidence before a House of Commons Committee appointed to report on the working of the first Post Office steam packets running between Holyhead and Dublin. The recommendations made to the P. and O. Company had some influence on the treatment of naval boilers, but in the early 'seventies the position was so serious, owing to the rapid corrosion, that the Admiralty in 1874 appointed a Committee to thoroughly investigate the matter.

Beginning their work in June 1874, for four years the Committee carried on their investigations into the properties of iron and steel; the construction, treatment, and preservation of boilers; the effects of salt and fresh water; and the use of lime, soda, boiler compositions, and zinc slabs. The inquiries were not confined to naval practice, but included practice ashore and in mercantile ships. In 1878 the original Committee was superseded, and a "limited" Committee was appointed to complete the work. The results of the labours of these Committees are to be found in three voluminous reports which contain the evidence of more than 170 witnesses and many lengthy appendices. Many theories were advanced to account for the corrosion of boilers, and many remedies were suggested. The reports proved of great benefit to both the Navy and the mercantile marine, and the measures subsequently adopted, which included the extensive use of zinc slabs in boilers, brought about a great improvement. It was this long and exhaustive inquiry which led the Admiralty in 1879 to issue the first edition of that useful work, the *Steam Manual for Her Majesty's Fleet; containing Regulations and Instructions relating to the Machinery of Her Majesty's Ships*. Previously, instructions had been promulgated by circular letters and then incorporated in the *Queen's Regulations*. Of such circulars dealing with boilers one of the most important was that issued on August 4, 1874, and it is instructive to compare the orders given in that circular with those contained in the *Steam Manual* five years later.

It was while the Admiralty Boiler Committee was carrying out its investigations that the most disastrous boiler explosion in the annals of the Navy occurred. The accident took place in the battleship *Thunderer* on July 14, 1876, and it led to the death of more than 40 persons. The *Thunderer* had been supplied with eight box boilers constructed by Humphrys, Tennant and Company for a working pressure of 30 lb. per square inch. On the day of the accident, the ship proceeded out of Portsmouth Harbour to Stokes Bay to carry out a full-power trial. Many officials were on board to watch the trials, and the ship was just being put on her course, when without any warning, the upper part of the front of one of the boilers was blown out and the stokeholds and engine rooms were immediately filled with the released steam. Owing to the extraordinary nature of the explosion, the inquiry which followed was very thorough, and many eminent engineers gave evidence. The reasons for the explosion, however, were not far to seek, and they had nothing to do with the construction or condition of the boiler. It was shown conclusively that no one had opened the stop valves, and that the pressure gauge, having been found out of order, had been shut off, while examination and experiments with the safety valves in the shop left no doubt that they had become inoperative. The safety valves were of the deadweight type, and while the valves and seatings were of brass, the safety-valve box was a large iron casting. Had the stop valves been open, or the safety valves been in working order, the accident could not have occurred, while, if the pressure-gauge had been in use, attention doubtless would have been called to the undue rise in pressure. As it was, no one could say at what pressure the boiler gave way. For the inquiry, a model of the boiler as it appeared after the accident was made in Portsmouth Dockyard, and this is now preserved in the Science Museum.

The *Thunderer*, which was launched in 1872 and completed in 1876, was the last capital ship in the Navy to have rectangular boilers. While such boilers had proved satisfactory for steam pressures up to 30 lb. per square inch, as soon as it was decided to use higher pressures, recourse was had to boilers with cylindrical shells and furnaces. Cylindrical boilers had been in use on land for many years, and the change from the return-tube and through-tube box boiler to the return-tube and through-tube cylindrical or "Scotch" boiler presented no difficulty. With the change came the introduction of double-ended return-tube boilers. The first of this type are said to have been fitted in the *McGregor Laird* in 1862, the boilers of which were 10 ft. in diameter and 14 ft. 6 in. long, with two furnaces 3 ft. in diameter at each end, and a common combustion chamber. From that time onwards cylindrical boilers of one type or another became common. A sketch of a cylindrical marine type boiler is given in Fig. 26. In some ships, the boilers, instead of being circular in section, were oval, having semi-circular tops and bottoms, with straight sides, which were strengthened with angle iron and stay bars (see Fig. 23). The immediate successors of the *Thunderer*, H.M.SS. *Dreadnought* and *Inflexible*, had oval boilers, but the type was not extensively used. Although the change to the cylindrical form of boiler was dictated by necessity, it had its disadvantages, for it was found that, owing to the reduced size of the furnaces, cylindrical boilers were inferior to box boilers, both in evaporative power and economy in production of steam. Sennett, referring to this in the first edition of his work, *The Marine Steam Engine*, 1882, said that whereas in box boilers about 30 lb. of coal was burnt per hour and 10 indicated horse-power was developed per square foot of grate, in cylindrical boilers the corresponding figures were only 21 lb. and $8\frac{1}{2}$ indicated horse-power. Another disadvantage was that the furnaces were not all on the same level as they were in the box boiler, and therefore not so easily fired.

The next important improvement in marine boilers, after the adoption of cylindrical shells and furnaces, was the use of steel instead of iron. It has already been stated that steel was used to some extent in the 'sixties even for marine boilers. But it was not until the introduction of the open-hearth process, made possible by the invention of the regenerative furnace, that steel for boilers made much headway. The invention of the regenerative furnace was due to Friedrich Siemens (1826–1904), but its development was mainly the work of his brother William (afterwards Sir William) Siemens (1823–83), who spent the greater part of his life in

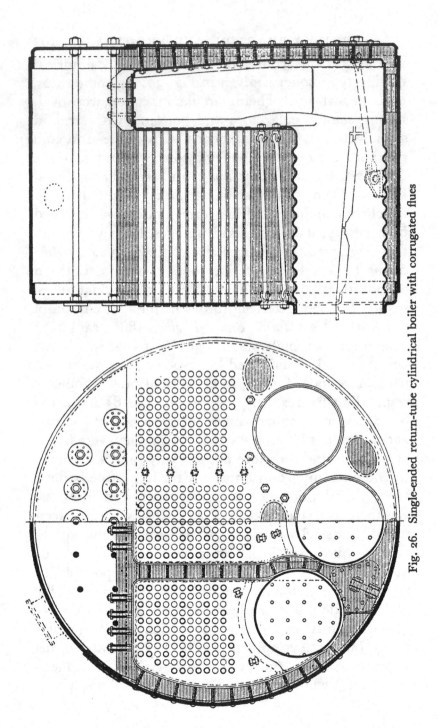

Fig. 26. Single-ended return-tube cylindrical boiler with corrugated flues

England. First applied to steel making in 1862, the regenerative open-hearth process was used at the Siemens Sample-Steel Works at Birmingham in 1865, and the success of this enterprise led to the founding of larger works at Landore near Swansea. During 1874–76 the Admiralty Boiler Committee addressed a questionnaire to many marine engineering firms and makers of iron and steel, asking for their views on the question of steel for boilers. Of the iron and steel makers, the Butterley Company in 1874 said that they had just installed a Siemens gas furnace, while the Landore Siemens Steel Company, replied on March 1, 1876, saying that they had been making steel for about twelve months and that in their opinion, "steel of the special quality we are now making will eventually supersede iron, and will be used for every part of the boiler". Few forecasts have been more completely justified. The Landore steel had an ultimate tensile strength of 30 tons, as against the 22 tons of wrought iron, and, moreover, it was homogeneous. It was this steel which was specified for the boilers of the *Iris* and *Mercury* in 1877, a year which was also marked by another important innovation in marine boilers, for on March 19, Samson Fox (1838–1903) patented the corrugated flue shown in Fig. 26.

While the great majority of ships, both naval and mercantile, were fitted with either box boilers or cylindrical boilers, quite a number of vessels were supplied with water-tube boilers. Unfortunately few of these early installations proved successful. An account of some of the experiments made in British ships is on record in a paper by Sir Fortescue Flannery, entitled "The Construction of High Pressure Boilers", read before the Institution of Civil Engineers on May 7, 1878, while the important work done in France was reviewed in Bertin and Robertson's *Marine Boilers*, 1898. A water-tube boiler consists essentially of a vessel into which the feed water is pumped, tubes exposed to the fire and gases, and through which the water and steam flow, and lastly a receiver from which the steam is withdrawn. These three elements have been used in an amazing variety of forms. As for the tubes alone, they have been used in coils, round or flattened, vertical or horizontal; in straight lengths; in lengths slightly curved or elaborately bent and they have been placed at every possible angle. Water-tube boilers were in use more than a century ago and their history includes the story of the coil boilers used by Rumsey and Fitch, the American pioneers of the steam boat.

Bertin divided water-tube boilers into three categories, viz., boilers with limited circulation, those with free circulation, and those with accelerated circulation. In the last category he placed such well-known boilers as the Yarrow, Thornycroft, Normand and Du Temple. All these boilers have two lower water drums and one upper steam drum connected by straight or curved tubes. A valuable review of the early history of boilers with accelerated circulation is contained in the two works of M. Paul-Augustin Normand, *Les Origines des chaudières à circulation accélérée (1825–85)* and *Quelques autres précurseurs des chaudières à circulation accélérée*, published first in the *Revue Maritime* and then separately in 1929 and 1931 respectively. In these it is shown that the three-drum type of boiler with straight tubes, like the Yarrow, was first patented by Clark and Motley in 1849, while a three-drum type with curved tubes was included in the designs of the Rowans, father and son, who were the pioneers in the use of water-tube boilers in the British mercantile marine. J. M. Rowan was the owner of the Atlas Works at Glasgow, and the first ship fitted by him with a water-tube boiler was the *Thetis*, built by Scott of Greenock in 1858. The working pressure was no less than 115 lb. per square inch, and as the ship had a compound engine, the whole installation was in the nature of a bold experiment. On a trial conducted by Rankine, the coal consumption was found to be only 1·02 lb. per horse-power per hour. The ship ran successfully for a time, but trouble then occurred with the boilers and they were removed. With his partner, T. R. Horton, and his son, F. J. Rowan, the elder Rowan continued to develop the water-tube boiler for marine purposes, and the firm subsequently supplied boilers for river boats in India. The most notable installation of the Rowan boilers afloat was that in the *Propontis*, a vessel which also had a triple-expansion engine designed by A. C. Kirk. But this pioneering effort was no more successful than that of the *Thetis*. Two serious boiler explosions occurred, resulting in the decision to revert to boilers of a more conventional type, and it remained for others to succeed where the Rowans had failed.

Of the boilers with limited circulation, the most famous is that of the French engineer, Julien Belleville, whose boilers were tried about the same time as Rowan's. At first Belleville used vertical iron coils, but trials carried out in the *Biche* (1856) and *Argus* and *Sainte Barbe* (1861) proved unsuccessful. Ten years later the Belleville boiler reappeared with the tubes horizontal, and from that type was evolved the boiler with zigzags, composed of slightly inclined tubes connected at the ends with horizontal junction boxes, as first seen in the high-speed boat *Hirondelle* in 1872. In these boilers automatic-feed regulators were adopted, and subsequent improvements included the use of downcomers and mud drums. Tried again in the *Voltigeur* in 1880, four years later the Belleville boiler was adopted for the Messageries Maritimes, and it was subsequently used for many of the most important warships.

In the 'seventies, no water-tube boiler attracted more attention in England than that of Loftus Perkins, in which steam was generated at several hundred pounds pressure. As early as 1823, Jacob Perkins (1766–1849) began experimenting with high-pressure steam, and his son Angier March Perkins (1800–81) and his grandson Loftus Perkins (1834–91) were both inventors of steam plant. A Perkins boiler outwardly resembled an ordinary vertical boiler. The heating surfaces consisted of 16 elements, each having 11 horizontal iron tubes, 3 in. diameter, connected together vertically by short lengths of iron tube of $1\frac{5}{16}$ in. diameter. The tubes were tested hydraulically to 4000 lb. per square inch, and the whole boiler to 2000 lb. The working pressure in such boilers was often as high as 300 lb. to 500 lb., and after examining the Perkins system, the Admiralty Boiler Committee recommended that it should be tried in the Navy. Orders were therefore given for machinery of the Perkins type to be fitted in the *Pelican*, but differences arose between the contractors, the Yorkshire Engine Company, and the inventor, and after the boilers and engine were partly completed, the scheme was abandoned. Between 1878 and 1880, however, Perkins water-tube boilers were fitted in the privately-owned vessels *Wanderer*, *Irishman*, and *Anthracite*, of which some particulars were given in a paper entitled "Historical References to the Progress in the Use of High-Pressure Steam", read before the Institution of

Engineers and Shipbuilders in Scotland, on June 21, 1927, by J. Mollison. Some of the early marine water-tube boilers had failed through defective circulation, others through structural defects, but the failure of the machinery of the *Wanderer* Mollison attributed to the want of efficient feed pumps. The *Anthracite*, it may be added, crossed the Atlantic in 1880, using steam at 350 lb. pressure.

While Rowan, Belleville, Perkins and others were struggling with the problem of the water-tube boiler, other inventors were experimenting with various methods for accelerating the combustion in marine boilers. To follow the practice of locomotive engineers, and use the exhaust steam for creating a draught was impracticable, and the use of steam jets in the base of the funnel proved very extravagant. Fans and blowers had been tried by some of the pioneers, and in America, Edwin Augustus Stevens, in 1827, fitted the boilers of the *North America* with closed ashpits into which the air for combustion was forced by fans. Afterwards, in conjunction with his brother, Robert Livingston Stevens, he tried fans placed in the base of the funnel, and also fans forcing air into a closed stokehold. He was thus a pioneer of the three systems known as closed-ashpit forced draught, induced draught, and closed-stokehold forced draught. But like many other inventors, he was before his time, and it was not till the era of the compound engine, and the demand for high speed, especially in torpedo boats, that forced draught made much progress. With the development of torpedo craft we shall deal in a later chapter, but it may be recalled here that the history of such vessels really begins with H.M.S. *Lightning*, built in 1877, at Chiswick, by Thornycroft. A year previously, Thornycroft had built the yacht *Gitana*, and for both vessels he used locomotive boilers fitted in closed stokeholds into which air was forced by a steam-driven fan. In the *Lightning* he was able to burn as much as 100 lb. of coal per hour per square foot of grate. About the same time, the closed-stokehold system was adopted in France; in 1879 it was applied to the Chinese cruisers *Yang Wei* and *Chow Yung*, built by Hawthorn, Leslie and Company, and at the same time the torpedo ram *Polyphemus* was fitted with locomotive boilers in closed stokeholds. The next installations of the system in the Navy were those in the sloop *Satellite* and the turret-ship *Conqueror*, the trials of which formed the subject of a paper by R. J. Butler, read before the Institution of Naval Architects in 1883. From thence onward practically all naval vessels were fitted with closed-stokehold forced draught.

The main object in using forced draught in warships was to obtain the greatest possible power from the machinery when high speed was called for. In mercantile vessels forced draught was employed solely for increasing the efficiency and economy, and the system which has achieved the greatest success is that invented by James Howden (1832–1918).

Quite early in his career Howden turned his attention to the use of compound engines, and he embodied in one of his patents methods of supplying hot air to the furnaces. His earliest experiments with forced draught were made in 1862, but it was not till 1880 that he evolved the plan now known by his name, and in which the air for combustion is heated by the waste gases, and then forced into closed ashpits. The system was successfully applied in 1884 to the *New York City*, a vessel with compound engines working with steam at

80 lb. pressure. As with other innovations, there were difficulties to be overcome, but the Howden system soon acquired a wide popularity. Unlike its rival system, it never had to face the objections of persons who declared that the closed-stokehold system would have a bad moral effect on the stokers!

CHAPTER XIV

# INTRODUCTION OF AUXILIARY
# MACHINERY I

S O far, only the history of the propelling machinery of ships has been considered. The application of engineering to ships, however, was not confined to their propulsion, but was extended to many of the operations on board which before had been performed by manual labour. Especially was this the case in men-of-war, and before the end of the 'seventies of last century the capital ship had become, as the eminent naval architect, Mr (afterwards Sir) Edward Reed said, "a steam being". In a letter on Naval Administration, published in *The Times*, of January 19, 1877, Reed wrote:

Every war vessel is now a steamer, and some of our most powerful and valuable ships have not a sail upon them; but, on the contrary, are huge Engines of War, animated and put into activity in every part by steam and steam alone. The main propelling engines are worked by steam, a separate steam engine starts and stops them; steam ventilates the monster, steam weighs the anchors, steam steers her, steam pumps her out if she leaks, steam loads the gun, steam trains it, steam elevates or depresses it. The ship is a steam being...."

The ships Reed, no doubt, had in mind at the time, were the *Thunderer*, *Dreadnought* and *Inflexible*, the last of which was launched in 1876, and was completed two years later. When describing the trials of the *Inflexible*,

*The Times* referred to the extended use of steam machinery in her, and said:

Besides the main engines above described, this immense floating machine contains the following auxiliary engines: A horizontal direct-acting steering engine (Forrester's patent); two vertical direct-acting fire engines; a capstan engine (Harfield's patent); a small vertical direct-acting turning engine; two vertical direct-acting donkey engines for pumping out the bilges; four auxiliary feed engines, of the same character as the donkeys, and placed one in each stokehole; four of Brotherhood's patent three-cylinder fan engines; a couple of horizontal direct-acting centrifugal engines for circulating the water through the condensers; two reversing engines (steam and hydraulic power combined); two pairs of steam and hydraulic engines for working the shot hoists and the gun-loading apparatus, and turning the turrets; four vertical direct-acting ash-hoists; a couple of 40-horse-power engines for pumping out the main drains; two steam shot hoists; two of Brotherhood's three-cylinder engines for lifting and lowering boats; and four of Friedman's patent ejectors. Altogether, then, the *Inflexible*, when commissioned, will have no fewer than 39 engines aboard.

The first purpose for which separate auxiliary steam engines were used aboard ship was for feeding the boilers when steam was up and the main engines were stopped. These so-called "donkey" engines came into use with the adoption of tubular boilers. Most of these donkey pumps were fitted with crank and flywheel, although Penn in the 'forties introduced a single-cylinder horizontal direct-acting pump without a crank. Bourne in the first edition of his *Treatise on the Steam Engine*, 1846, gave a description and sketch of this pump, and also of an automatic regulating device used with it for maintaining the water in the boiler at the proper height. The credit for the invention of the direct-acting steam pump of this type is generally given to the American engineer Henry Rossiter Worthington (1817–80), who saw the need for an auxiliary steam pump in canal boats, and who took out his first patent for a pump in 1841. Some of his pumps were controlled by floats within the boiler "but", he said, "they proved dan-

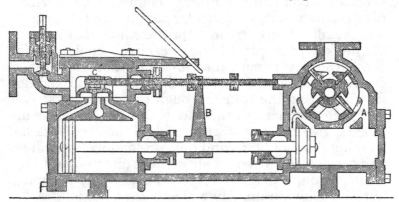

Fig. 27. Worthington direct-acting pump, 1849

gerous just as in proportion as they seemed reliable, because they did so well that it made the firemen careless and inattentive". Worthington tried various valve arrangements, but in 1849, in conjunction with W. H. Baker, he took out a patent for a direct-acting steam pump with a simple ordinary slide valve, and this was followed by the invention of the direct-acting duplex pump. The drawing of Worthington's pump as described in his patent is given in Fig. 27. Many types of direct-acting pumps have been used on ships at various times, but in the middle of last century most marine engineers preferred those of the crank-and-flywheel type.

For clearing the bilges of ships of water, hand pumps had been in use for centuries, and they continued to be relied upon long after the advent of steam. Experience soon showed, however, that the main propelling engine could, on occasion, be used for clearing the ship of water. In an emergency the snifting valve on the condenser could be opened, thus enabling the air pump to draw water from a flooded bilge, and later on a proper connection was fitted for the purpose. In an Admiralty specification of 1845, for a set of machinery for a paddle-wheel vessel, it was required that a small donkey engine should be fitted for feeding the boiler, that bilge injection pipes should be fitted to the condenser, and that a hand pump should be supplied, capable of being worked by the engine, and fitted with valves for drawing from the boilers, the bilge or the sea, and for delivering into the boilers, on deck or overboard. One early instance in which the main engine was used for attempting to clear the ship of water was that of the paddle vessel H.M.S. *Thunderbolt*, which was wrecked off the coast of South Africa on February 3, 1847; while a later one was that of H.M.S. *Royal Albert*, which, in December 1855, had to be beached on the Island of Zea, while repairs were effected to the stern gland.

The earliest form of marine engine always included an air pump and generally a feed pump, and to these were subsequently added a bilge pump and, in a surface condensing engine, a circulating pump. These were always either lift pumps or force pumps. It was in the 'seventies that main-feed pumps driven by separate engines began to be used, and about the same time separately driven centrifugal pumps began to be used as circulating pumps. Centrifugal pumps had been designed as early as 1680, but the first really successful pump of this type appears to have been set up at Boston, Massachusetts in 1818. Later on, many improvements in design were made, notably by the American inventor Andrews, who, in 1839, brought out the volute casing; by John George Appold (1800–65), who determined by experiments the most efficient curve for the blades of the impeller; and by Professor James Thomson (1822–92), who, in 1850, suggested the whirlpool chamber—a space surrounding the wheel in which a free vortex could be formed. An Appold pump gained a Council medal at the Great Exhibition in 1851, and it is now preserved at King's College, London. When first used aboard ship centrifugal circulating pumps were sometimes driven by gearing from the main engine, but this plan was soon abandoned in favour of separately driven pumps, the first of which, according to John Bourne, was fitted by himself in the *Jumna*. The facility with which centrifugal pumps dealt with large quantities of water made them especially valuable in ships, and in a few instances the pump casing was placed horizontal, low down in the engine-room, with the impeller driven by a vertical shaft from an engine placed on the engine-room gratings, so that the pump could still be worked even if the engine-room were partially flooded. In the *Inflexible* the engines of the pumps were placed at so high a level in the vessel that they could still be worked when there was 12 ft. of water in the engine room. The total pumping capacity of the *Inflexible* was 4800 tons per hour. Centrifugal pumps were also adopted for important mercantile vessels, the Cunard liner *Servia*, 1881, having, for instance, three pumps by J. and H. Gwynne, two of which had a combined capacity of 6000 tons an hour.

For fire purposes, when steam was not up, ships had to rely on the hand pumps. The inadequacy of these led the Admiralty, about 1880, to supply one or two capital ships, among them the *Sultan*, with fire pumps made by the well-known fire-engine makers, Shand, Mason and Company. In the *Sultan* the boiler was placed on the lower deck, and close beside it was a horizontal fire engine discharging into the fire main and also a three-throw pumping engine for drawing from the main drain. At a trial of the plant on January 6, 1880, steam was raised in ten minutes to 100 lb. per square inch, and a stream of water discharged over the mast 200 ft. above the level of the engine. The capacity of the fire pump was 1120 gallons per minute, and that of the pumping engine 720 tons per hour, or about eight times the capacity of all the hand pumps together. Another innovation of the time was the fitting of several large Friedmann steam ejectors for dealing with large quantities of water in case of emergency.

While the problems of applying steam to pumps aboard ship differed little from those associated with pumping in general, the proposal to apply steam to the steering of ships was entirely novel, and by many was considered an impracticable one. With the great increase in the size of ships, steering by manual labour, however, had become increasingly difficult and dangerous, and the need for some other method was urgent. Sometimes nearly a hundred men had to be stationed at the wheels of large steam vessels, and in stormy weather their work was extremely exhausting. There were, moreover, many accidents, and men were actually killed while trying to control the wheels. In H.M.S. *Minotaur* 78 men were required at the wheels, while it took one and a half minutes to put the helm over 25°. The pioneer of the steam steering engine was Frederick Elsworth Sickells, who was born in 1819 and died at

Kansas City, March 9, 1895. In 1859 he obtained a British patent for his engine; in 1862 he exhibited an engine at the International Exhibition in London, and about the same time he fitted a Federal steamer with steam steering gear. He included in his invention the overtaking valve gear which causes the engine to run only so long as the steering wheel is being turned. Unfortunately, his work attracted little attention, and lacking business ability and perseverance, he failed to proceed with his invention, and it remained for the Scottish engineer, John Macfarlane Gray (1832–1908), to design the first successful steam steering engine and to inaugurate the era of steering by steam. Gray's engine was made in 1866 and fitted in the famous *Great Eastern*, and as a piece of invention to meet an urgent practical need it may well be placed alongside the invention of the steam hammer by James Nasmyth.

Macfarlane Gray, who became widely known for his researches on thermodynamics and as the chief examiner of engineers under the Board of Trade, in 1866 was manager and chief draughtsman of the firm of George Forrester and Company, Vauxhall Works, Liverpool. His invention was fully described in a paper entitled "Description of the Steam Steering Engine in the *Great Eastern* Steamship", which he read before the Institution of Mechanical Engineers, on October 31, 1867. It is there stated that the conditions which had to be fulfilled were:

1. The alterations on board the vessel to consist of additions only, and no existing appliance for steering to be removed or impaired.

2. The steam power to be sufficient for moving the rudder at all times at a greater speed than had ever been attained by manual force.

3. The steering to be controlled from the bridge by some means equal in efficiency to the existing telegraph plan.

4. The new arrangements to admit of men working at the wheels in order to assist the steam power if required.

5. The steam gear to be capable of being disconnected without interfering for an instant with the means of steering by hand.

6. The rudder to be held in any position by the steam gear without applying a brake.

7. The engine to start without hesitation at all times, even after standing for a considerable period without acting.

8. The apparatus to be simple in all its parts and requiring no special attendant or engineer.

9. The engine to be capable of being used as a windlass if required, when disconnected.

One of the drawings accompanying Macfarlane Gray's paper is shown in Fig. 28.

The *Great Eastern* in 1866, it may be recalled, had already been used for cable laying and was fitted with a considerable amount of machinery for this purpose, but none of this was of a more ingenious character than Gray's steering engine which so successfully fulfilled the onerous conditions laid down. The engine, designed to work at 20 lb. pressure, had two horizontal cylinders, the pistons driving cranks at right angles. Each cylinder had a slide valve with no lap, and the reversing and control were effected by a differential valve and screw hunting gear in the manner almost universally adopted since. The hand steering wheel on the bridge was connected to the engine by 410 ft. of shafting of $1\frac{1}{4}$ in. solid-drawn iron tube, carried in some fifty cast-iron bearings fixed in one of the box beams which formed

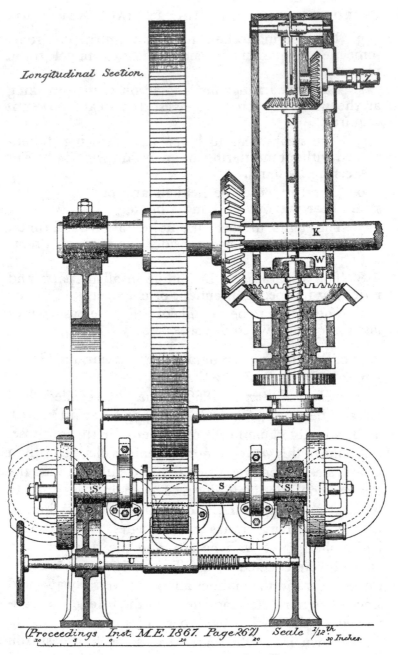

Longitudinal Section.

(Proceedings Inst. M.E. 1867. Page 267.)   Scale $\frac{1}{12}$th Inches.

Fig. 28.  Steering engine of *Great Eastern*

part of the upper deck. The backlash in the gear was only about one-eighth of a turn. The first trial of the engine took place in March 1867, and was entirely successful. Yet it was still thought by the commanding officer, Captain (afterwards Sir) James Anderson, that it would be advisable to steer by hand when in sight of land and in storms, and he left for America with this idea. His voyage more than dispelled any doubts he entertained, and he wrote from New York: "The steam steering gear seems perfection."

The fitting of the steam steering gear to the *Great Eastern* was followed by trials in other ships of more than one system of controlling tillers by hydraulic rams. One of these was due to Captain (afterwards Admiral Sir) E. A. Inglefield (1820–94) who described his system in a paper entitled "On the Hydraulic Steering Gear as being fitted to H.M.S. *Achilles*", read before the Institution of Naval Architects, on March 19, 1869. The *Achilles* was a battleship of 6121 tons, launched in 1863. Her engines were of 5600 horse-power, but she carried masts and yards. As it was considered that steam power would not always be available, Inglefield utilised the pressure due to the head of sea water near the bottom of the ship to obtain the necessary power. Low down in the hull was a large cylinder and piston to which the sea water had access. The piston rod extended in both directions so as to form small hydraulic plungers giving a pressure of about 600 lb. per square inch. This pressure-water was conveyed by pipes to two hydraulic cylinders connected to a Rapson slide actuating the tiller. The helmsman worked a slide, the water pressure pushed the helm over and, the slide being centred, the helm was locked in position. The used sea water from the working cylinder in the hold was allowed to escape into the bilge, whence it was removed at intervals by the ordinary pumping arrangements of the vessel. In another hydraulic system, the pressure was supplied from steam pumps driven by steam from a donkey boiler, but all such arrangements soon gave place to steam steering gears. All steam gears were not as successful as Gray's in the *Great Eastern*, and as late as 1892, a writer in *Brassey's Annual* said "The man who shall invent a system of steam steering that shall be as trustworthy as, say, the propelling engines of an average man-of-war, will deserve well of his country. At present no such system exists."

Just as the increase in the size of ships led to the application of power for steering, so the increase in the size of naval guns led to the introduction of machinery for controlling them. A sketch of the history of the development of guns through the work of Armstrong, Whitworth and others is given in Robertson's *The Evolution of Naval Armament*. With the construction in France of the *Gloire*, and the *Warrior* in England, had begun the never-ceasing duel between guns and armour, while, in the *Monitor*, Ericsson had shown how the largest guns could be mounted in turrets and trained by steam. It is difficult, says Robertson to trace to its source the invention of the armoured gun-turret, but the turret idea, like so many other inventions, had an independent development in Europe and America. According to Chief Engineer J. W. King, U.S.N., who, in 1881, published a valuable work entitled *The War-Ships and Navies of the World*, Captain Eads of Saint Louis, U.S.A., invented as early as 1861, and soon afterwards successfully applied to two gunboats, a system of mounting heavy guns on a turntable within a rotating turret. The table, with its guns and attachments, was raised and lowered and revolved by steam power. The guns were also moved out to the firing position by the same medium, and the recoil was taken by steam pressure. Eads subsequently invented and patented the principle of raising and lowering guns by the elastic force of compressed air, a system similar to that developed in England by Colonel Moncrieff.

In our own country, the foremost advocate of the turret system was Captain Cowper Phipps Coles (1819–70), who perished in the *Captain* disaster of September 7, 1870. In 1855, when in command of H.M.S. *Stromboli* in the Sea of Azov, Coles constructed a raft capable of bearing heavy artillery, and the same year sketched a design for a shallow draught vessel carrying guns protected by a fixed hemispherical iron shield. From this it was but a step to the cupola or turret on the centre line of the ship, and in 1863 he was enabled to put his ideas into practice in the *Royal Sovereign*, a wooden three-decker of 3144 tons, carrying 120 guns, with engines of 800 horse-power. For the purpose, the two uppermost decks were removed, and the ship was strengthened with heavy timbers and iron plating and armour. "From the side of the ship", said one writer, "the deck sloped upwards to the outer circumference of the turrets, which stood like so many circular revolving forts on the crest of a glacis." The armament consisted of five $12\frac{1}{2}$-ton guns mounted in four turrets. In the American monitors the base of the turret, when not in action, rested on a brass ring, and before it could be moved its weight had to be transferred to a central vertical spindle by means of a screw wedge. In the *Royal Sovereign* the weight of each turret was distributed on a ring of bevelled rollers similar to those used in a railway turntable, and in the manner common to-day.

In the centre of each turret there was also a vertical wrought-iron cylinder sufficiently large for men to pass through. In spite of the demonstration in the United States of the possibility of applying steam for the control of turrets, those of the *Royal Sovereign* were revolved by manual power by means of racks and pinions or by handspikes worked like capstan bars on the lower deck.

The first example of a turret worked by steam, seen in the British Navy, was that of the *Thunderer*, launched in 1872, and it was in her and her successors, the *Dreadnought* and *Inflexible*, that steam and hydraulic machinery were first applied extensively for the performance of all the main operations in the working of guns. The fore turret of the *Thunderer* (see Fig. 29), to which steam was first applied, was $31\frac{1}{4}$ ft. in diameter and weighed, with its guns, 406 tons. It was mounted on a ring of rollers as in the *Royal Sovereign* and around the base of the turret was an external toothed rack. The steam engine for driving the toothed pinion was placed below the turret, and it could be controlled both from inside and outside the turret. Both the turrets of the *Dreadnought*, which each carried two 38-ton guns, were worked in the same manner, and when reporting on the trials on October 15, 1878, *The Times* remarked that

The only defect is in the steam machinery which rotates the turrets, the elasticity of the motive power rendering it occasionally difficult to stop the guns precisely in the loading position, notwithstanding the ingenuity of the locking arrangement. In the *Inflexible*, however, the rotating as well as the loading gear will be worked by hydraulic power, a power which is completely under the command of the operator.

As is well known, the first application of hydraulic power to machinery was made by William George

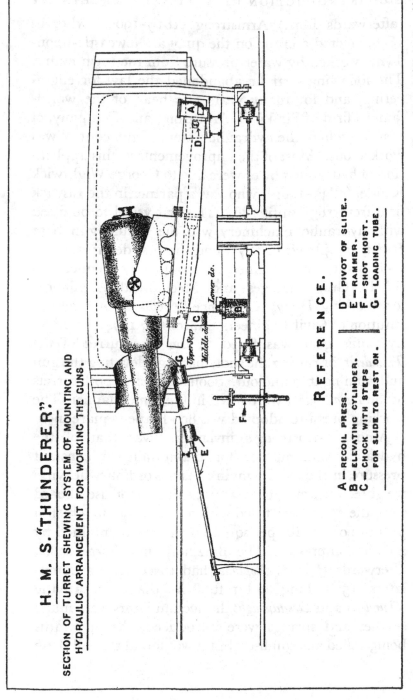

H. M. S. "THUNDERER".

SECTION OF TURRET SHEWING SYSTEM OF MOUNTING AND
HYDRAULIC ARRANGEMENT FOR WORKING THE GUNS.

REFERENCE.

A — RECOIL PRESS.
B — ELEVATING CYLINDER.
C — CHOCK WITH STEPS
       FOR SLIDE TO REST ON.
D — PIVOT OF SLIDE.
E — RAMMER.
F — SHOT HOIST.
G — LOADING TUBE.

Fig. 29.

From *Manual of Hydraulics*, 1883

(afterwards Lord) Armstrong (1819–1900), who, in 1846, erected a crane on the quay at Newcastle-upon-Tyne, worked by water pressure from the town mains. The following year he abandoned the law for engineering, and for many years was head of the world-famous firm of Sir W. G. Armstrong and Company, of Elswick, where the hydraulic system of gun control was worked out. Most of the improvements in this application of hydraulic power were due to George Wightwick Rendel (1833–1902), who was a partner in the Elswick firm from 1858 until 1882. The first vessel to be fitted with hydraulic machinery was the Dutch gun boat *Hydra*, completed in 1873, and the first descriptions of such plant are contained in a paper by Rendel, entitled "Gun-Carriages and Mechanical Appliances for Working Heavy Ordnance", read before the Institution of Civil Engineers, March 10, 1874. In 1875, hydraulic gear was fitted to the fore turret of the *Thunderer* for raising the shot, loading, elevating the gun and running it in and out. Soon afterwards both turrets of the *Dreadnought* were fitted in the same manner. The working pressure adopted was 800 lb. per square inch.

Among Armstrong's inventions was that of the hydraulic accumulator for maintaining a constant pressure in the supply mains. As the ordinary heavily-weighted accumulator was not suitable for use aboard ship, the *Hydra* was fitted with an air vessel in which a pressure of 800 lb. per square inch was maintained by an air compressor. The Brazilian ship *Independencia*, afterwards H.M.S. *Neptune*, had three such accumulators, 15 ft. long and 1 ft. 8 in. diameter. In the *Thunderer* and *Dreadnought* the accumulators were much smaller and springs were introduced, the apparatus being called an equaliser, but it was found that all these arrangements could be dispensed with by the use of an hydraulic-steam throttle governor, which enabled the pumping engine to maintain a steady pressure of 800 lb. per square inch, irrespective of the amount of pressure-water being used. At the same time, Rendel's three-cylinder hydraulic motor was adopted for rotating the turrets and both this and the new governor gear were fitted in the *Inflexible*, which had four 80-ton guns mounted in two turrets, each of which weighed, with its guns, 750 tons. At a trial it was found possible to turn a turret one complete revolution in 1 minute 8 seconds, and to have it completely under command when rotated at a creeping pace. These performances may be said to have marked the close of a stage in the improvements in the method of mounting and working guns, which began about twenty years earlier with the substitution of iron carriages for wooden carriages and of steam power for manual labour.

CHAPTER XV

# INTRODUCTION OF AUXILIARY
# MACHINERY II

THE same period which saw the application of steam to pumping, steering, and the working of guns in ships also saw the introduction of distilling apparatus, and of air-compressing machinery for charging torpedoes, while later on steam was used for driving electric generators and refrigerating plant. Nothing, perhaps, contributed more to the improvement of the health of the ships' companies in the Navy than the use of distilled water. Simple as the process of obtaining fresh water from sea water appears to-day, its introduction on any considerable scale did not take place till long after the introduction of steam propelling machinery. The invention of the process appears to have been made, tried and discarded several times. The earliest reference to distillation from sea water goes back to the third century A.D. Sailors, it says, "boil the sea water and suspend large sponges from the mouth of a brazen vessel to imbibe what is evaporated and, in drawing this off from the sponges, they find it to be sweet water". Some interesting notes on the subject are given by Admiral P. H. Colomb in his *Life of Admiral Sir Astley Cooper Key*:

The march of invention had been in nothing slower than in the vital necessity of water for ships. As is well known, Sir Richard Hawkins had on board his ship a distilling apparatus during his voyage to the South Sea in 1593, of which he thus writes: "with an invention I had in my shippe I easily drew out of the water of the sea sufficient quantitie of fresh water to sustain my people with little expence of fewell; for with foure billets I stilled a hogshead of water, and therewith dressed the meat for the sick and whole. The water so distilled we found to be wholesome and nourishing."

A proposal for distilling at sea is referred to in a letter preserved in the library of Worcester College, Oxford. The letter is signed by Pepys, and dated from Windsor, August 3, 1684, and it contains instructions to Captain William Gifford, of H.M.S. *Mermaid*, to receive on board and make experiments with an engine for "producing Fresh Water (at Sea) out of Salt", if it could be done without danger to the ship through fire.

Colomb also refers to the inventions of Dr Lind and Dr Irvine. Dr Lind, who was born in 1716 and died in 1794, was from 1758 till his death, surgeon-in-charge of Haslar Hospital. In 1761, he demonstrated before the Portsmouth Academy that the steam from salt water was fresh, and the following year showed his experiments to the Royal Society. A few years later Dr Irvine received a reward of £5000 from the Government for introducing distillation aboard ship, but although at the beginning of the nineteenth century several line-of-battleships, East Indiamen, and merchant vessels had "fire hearths" for distilling, the plan made little or no headway.

When steam boilers and engines were fitted in ships the problem of distilling became somewhat simpler. The practice of using water condensed in steam cylinder jackets has already been referred to. The same question of water supply occupied the attention of Mr (afterwards Sir) Thomas Tassell Grant, F.R.S. (1795–1859), who, while storekeeper of the Clarence Victualling Yard, Gosport, in 1834 brought out a method of distilling, which was adopted fairly extensively about fifteen years later. An interesting item in a naval machinery specification of 1845 stated that a small, flat iron vessel was to be fixed in one of the paddle-boxes with two pipes, one communicating with the stokehold and the other with the boiler, for obtaining a small supply of distilled water. In the *Life and Letters of Admiral Sir B. J. Sulivan* (1810–90) there is an account of the use of a somewhat similar device. During the Russian War of 1854–56 Sulivan was in command of H.M.S. *Lightning*. In this ship an exhaust-steam pipe led against the copper plate forming a part of the paddle-box, and the condensed water was caught in grooves and led into buckets. Sulivan noticed the engineers using this soft water for washing, and directed that it should be placed in tanks and aerated. The water so obtained afterwards formed the chief supply of drinking water for the crew, and to its use Sulivan attributed the immunity of the crew from the cholera which prevailed in the fleet. But, generally speaking, water for ships was still obtained from shore. "Casks were landed, rolled up the beach, filled and rolled down the beach, and 'parbuckled' into the launches", and then stowed aboard.

An early distilling condenser as fitted to ships in the Navy in the middle of last century, consisted of a nest of 154 vertical copper tubes, each 3 ft. 6 in. long and 1 in. internal diameter, expanded into an upper steam chamber and a lower water chamber, the whole being enclosed in a tank about 5 ft. by 4 ft. 6 in. by 2 ft., through which the sea water flowed by convection. Steam was supplied to the apparatus from one of the ship's boilers or an auxiliary boiler fed from the sea. The apparatus was placed low down in the hold and the distilled water was pumped into the ship's tanks by hand. An important improvement in distilling was made by Alphonse Normandy (1809–64), the French chemist who came to England in 1843, and in 1851 took out a patent for a method of distilling. The merit of his invention consisted of conducting the operation at a low temperature and causing the condensed water to absorb a large quantity of atmospheric air which rendered it palatable. In a Normandy distiller, there were two sections, one for condensing the steam and another for cooling the condensed water. His distiller is illustrated in Fig. 30.

The distillation of water for use in boilers was practically forced upon marine engineers by the introduction of triple-expansion engines working with steam at high pressures, and it led to the use of evaporators fitted with steam coils. The Frenchman, Sochet, had brought out a multiple-effect evaporator as early as 1832, and Samuel Hall, in his patent of 1834, for a surface condensing engine, had included a still or evaporator placed in the steam space of the main boiler. Dinnen, in 1838, had stated that nothing but the use of distilled water would thoroughly obviate the evils to which marine boilers were subjected on long voyages, but he found few supporters. Many strange views were held on the subject, and when the Admiralty Boiler Committee of 1874 was taking evidence, one naval engineer said that distilled water was considered an insidious thing that it was best to have nothing to do with! When distilling from steam produced in ships' boilers was in vogue, the distilled water was seldom really pure and this itself caused a certain amount of scepticism. Loftus Perkins, indeed, said to the Boiler Committee, "It seems that you cannot distil salt water. There is no distilling apparatus yet invented by which you can make pure water out of salt water. It is impossible to make pure distilled water at all I think; it is only comparatively pure...." Some of the difficulties at least were removed

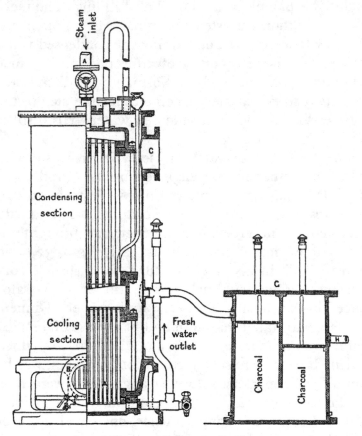

Fig. 30. Normandy distiller

with the reintroduction of evaporators fitted with steam heating coils. An evaporator of this type was fitted in the French gunboat *Crocodile* in 1873, and in the 'eighties the well-known evaporators of Weir, Caird and Rayner, Kirkaldy and others made their appearance. Since then successive improvements in evaporators and distillers have enabled all ships to be supplied with distilled water in abundance both for boilers and for drinking purposes.

Among the inventions which have materially affected the progress of naval engineering is that of the automobile torpedo. Of such torpedoes the most important is that of Whitehead, the adoption of which had far-reaching results. Invented in the 'sixties and adopted by all the most important naval powers in the 'seventies of last century, the introduction of the Whitehead torpedo was followed by the installation in warships of air-compressing machinery and electric generating plant, and by the construction of fast torpedo boats and a demand for increased speed in other classes of naval vessels. Robert Whitehead (1823–1905) was born in Lancashire, trained as a mechanical engineer in Manchester and after working at Marseilles, Milan and Trieste established engineering works at Fiume, where he designed and constructed engines for the Austrian Navy. In 1864, he was appealed to for assistance by Captain Lupuis, an Austrian naval officer, who had been attempting to construct a surface torpedo propelled and directed from a fixed base by means of ropes and guiding lines, a method somewhat similar to that afterwards developed by Louis Brennan (1852–1932). Considering that a torpedo as conceived by Captain Lupuis would have but a limited sphere of usefulness, Whitehead set out to construct a torpedo which would be propelled beneath the surface of the water by its own machinery. By 1866 he had constructed a cigar-shaped machine, 14 in. in diameter, weighing about 300 lb., carrying 18 lb. of dynamite, and driven by a motor supplied with compressed air at 700 lb. per square inch, stored in a chamber in the torpedo. Brought first to the notice of the Austrian Government, this torpedo was afterwards reported on favourably by a committee of British naval officers, and in 1870 the Admiralty ordered experiments with it to be carried out at Sheerness. Those in charge of the experiments reporting that in their opinion "any maritime nation failing to provide itself with submarine locomotive torpedoes would be neglecting a great source of power for offence and defence", the Government purchased the secret, and the right of manufacture of the Whitehead torpedo for £15,000; and arrangements were at once made for torpedo construction on a large scale. From that step came the erection of factories and stores, the inauguration of schools of instruction, and the birth of the great organisation within the Navy, which to-day deals with torpedoes and everything connected with them.

Like all other machines, the Whitehead torpedo since its inception has been subject to constant improvement, and there is little resemblance between the mechanism of the earliest and of the latest torpedoes. Some of the first important improvements were made by Whitehead himself. In 1868, he added the balance chamber mechanism for maintaining the torpedo at a constant depth, and in 1876 he brought out his ingenious servo-motor for working the horizontal rudders. Another notable improvement was the use of the Brotherhood engine in place of the compound oscillating engines first fitted. Peter Brotherhood (1838–1902), who had studied at King's College, London, and had worked in the locomotive department at Swindon and at Maudslays in Lambeth, commenced business for himself in Goswell road, London, in 1867, and in 1872 brought out his three-cylinder single-acting radial steam engine with one crank. The three cylinders were arranged at angles of 120° around the crank chamber. In 1873, the engine was shown at the Vienna Exhibition and shortly afterwards was adopted as an air motor for driving torpedoes, for which it proved eminently suitable. The effect of these improvements was to give torpedoes greater range and greater accuracy. The first torpedoes had a speed of 8 knots. By 1876, a speed of 18 knots for a distance of 600 yards had been attained; by 1884 the speed had been increased to 24 knots, while by 1889, for the first 1000 yards the speed was 28 to 29 knots, and the charge carried was 200 lb. of guncotton.

The use of air at a pressure of about 1000 lb. per square inch in torpedoes introduced a new problem in engineering, for the air compressors which had been used in connection with mining and tunnelling had worked at pressures of only 100 lb. or less. These machines were often of the "wet" type, in which the cooling water was injected into the cylinder. Machines of the wet type were, obviously, unsuitable for torpedo work, and machines of the "dry" type, in which the compressing cylinders were water-jacketed, were developed. Here again much credit was due to Brotherhood, who devoted himself to the development of light and compact air-compressing plant for torpedo-boats and ships, and most of the machines used to-day are of the Brotherhood type, in which the compressing is carried out in two or more stages. A sketch of a section through a Brotherhood two-stage compressor is given in Fig. 31. The air-compressing plant of a capital ship of the early 'eighties consisted of one or more steam-driven compressors capable of compressing 30 cubic feet of air to a pressure of 1500 lb. per square inch per hour, and a reservoir consisting of a nest of 50 steel tubes 3 in. in diameter and 6 ft. long, connected by unions and pipes. Since then working pressures have risen to 2500 lb. or 3000 lb. per square inch, and the reservoirs are now groups of gas cylinders.

The use of electric generating plant aboard ship like that of air compressors was due to the adoption of the torpedo as a fighting weapon. The first Whitehead tor-

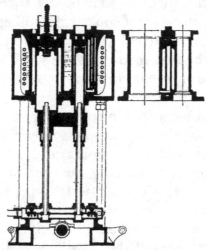

Fig. 31. Section through Brotherhood air compressor, 1877

pedoes were fired from tubes submerged beneath the surface of the water, but it was soon found possible to discharge them from above-water tubes, and also from specially constructed davits and dropping gear fitted to steam boats. Already steam launches had been used for attacking ships by means of outrigger or spar torpedoes, and it was the problem of defending ships at anchor from these attacks by small craft at night or in a mist which led to the use of searchlights. In 1873, Henry Wilde (1833–1919), the Manchester electrical engineer, had directed the attention of the Admiralty to the advantages of electric searchlights for naval purposes, and shortly afterwards he was enabled to carry out experiments at Spithead on board H.M.S. *Comet*. These proving satisfactory, in 1876 electric generators and searchlights were installed aboard the *Minotaur* and *Temeraire*, and in 1877 aboard the *Dreadnought*, *Neptune*, and other vessels. Soon afterwards searchlights were fitted to all ships. The history of the arc light, as used in searchlights, goes back to Davy's experiments in 1808, at the Royal Institution, but the first to produce a truly self-regulating arc lamp was William Petrie (1821–1908), who, during 1846–53, worked in collaboration with William Edward Staite at electric-lighting problems. In 1848, they exhibited a powerful arc lamp on the portico of the National Gallery, and the following year illuminated old Hungerford Bridge. Arc lights and reflectors were used by the French at the siege of Kinburn in 1855. Petrie's lamps were supplied with current from magneto-electric machines with permanent magnets, as were also the arc lamps installed in 1857 in the South Foreland lighthouse. By the 'seventies, however, the problem of the self-exciting dynamo with electro-magnets had been solved through the work of Siemens, Wheatstone, Varley, Wilde, Gramme and others, and it was this type of generator which was fitted in ships.

A further step in the use of electricity aboard ships was taken in 1879, when William Inman arranged for the lighting of the saloon of the liner *City of Berlin* by means of arc lamps; and when, a little later, Edison and Swan, working independently, solved the problem then spoken of as "the subdivision of the electric light", by the invention of the incandescent lamp, electric lighting of ships soon came into general favour. Both the British battleship *Inflexible* and the Inman liner *City of Rich-*

*mond* were fitted with Swan lamps in 1881, and about the same time the Cunard Company installed the electric light in the *Servia*.

By 1884, electric light had been installed in more than 150 ships. Much information regarding the early installation in both warships and merchant vessels is contained in the two papers, "Electric Lighting for Steamships", read by A. Jamieson, before the Institution of Civil Engineers, on November 11, 1884, and "On the Applications of Electricity in the Royal Dockyards and Navy", read by H. E. Deadman, before the Institution of Mechanical Engineers, on July 26, 1892. Everyone connected with ships appreciated the change from oil and candles to electricity, but none more so than those in the engine-room, for, as Alexander Siemens remarked, "it was now possible, for the first time, for marine engineers to see the engines working".

The electric generator fitted in the *Minotaur* was an alternator having 32 magnets, but those fitted in other ships of the Navy were direct-current machines by either Siemens, Edison, Gramme, Crompton, Wilde, or other well-known makers. The lighting installation in H.M.S. *Inflexible* was carried out by the Anglo-American Brush Company. It was a combined system of arc and glow lamps, arc lamps being fitted in the engine-room. The dynamos were of the Brush type, and each was capable of lighting 16 Brush arc-lamps of 2000 candle-power each. A Gramme dynamo supplied current to the searchlights. As in early electric power stations ashore, the dynamos in ships were, as a rule, at first driven by slow-running steam engines through gearing, belting, or cotton-rope drives. The generator in the *Minotaur* was actually driven by belting from an auxiliary steam pump. In 1876, a Brotherhood three-cylinder engine was directly connected to a dynamo fitted aboard the French battleship *Richelieu*, and subsequently many hundreds of similar lighting sets were built. The example of Brotherhood was followed by Westinghouse, Ferranti, Tangye and others, while in the later 'eighties there was no more popular engine for driving dynamos direct than the high-speed engine of Willans. The first ship in the United States Navy to have electric light, U.S.S. *Trenton*, was supplied in 1883 with a direct-driven Edison shunt-wound dynamo with an output of 120 amperes at 110 volts, the ship being fitted with 173 10-candle-power, 70 16-candle-power, and four 32-candle-power lamps. When the *Trenton* was lost in the hurricane at Samoa, March 16–17, 1889, the dynamo continued running until the fires in the boilers were extinguished by the water rising in the ship.

A lighting set in a British battleship in 1885 had an output of 150 amperes at 80 volts, the speed of the engine being about 300 revolutions per minute. A few years later a design was got out at Portsmouth Dockyard for a set with an output of 400 amperes at 80 volts, the dynamo being driven by a vertical compound engine. The first sets to this design were constructed in the Dockyard, but many similar sets were made by Brotherhood, William Henry Allen (1844–1926), and George Edward Belliss (1838–1909) of Birmingham, all of whom made improvements in marine auxiliary machinery. Belliss was first known in the Navy as a constructor of machinery for steam boats, but it is not without historical interest to recall that it was in connection with the construction of his electric-light engines that the present system of forced lubrication was introduced. The first patent for forced lubrication was taken out in 1890 by Albert Charles Pain (1856–1929), who had been a draughtsman in Portsmouth Dockyard, but had gone to Birmingham shortly after Alfred Morcom (1847–1905), the naval engineer, had resigned his position as Chief Engineer of Sheerness Dockyard, to join Belliss in partnership. A copy of the sketch given in Pain's patent is shown in Fig. 32. Further information regarding the application of electricity in the Royal Navy is given in a paper entitled "The Applications of Electricity in Warships", read before the Institution of Electrical Engineers on March 21, 1927, by Mr W. McClelland, the Director of Electrical Engineering at the Admiralty.

The last type of auxiliary plant introduced at this time was the refrigerating machine, fitted to enable frozen meat to be brought from distant countries. There have been many pioneers of mechanical refrigeration, including Jacob Perkins, who in 1834 invented a machine in which a vapour was compressed, condensed and evaporated, and Dr James Gorrie (1803–55) of Florida, who in 1850 patented a cold air machine. In 1876 Charles Tellier (1827–1913) fitted a vapour machine in the *Le Frigorifique* which sailed between France and the Argentine, and in the same year Carl von Linde (1842–1934) patented an ammonia machine. Other cold air machines were brought out by Henri Giffard, the Bell brothers and Joseph Coleman (1838–88), and Franz Windhausen devised a carbon dioxide machine. The first important refrigerating plant aboard ship was that in the *Strathleven* which left Melbourne on November 29, 1879, and reached London on February 6, 1880, having 34 tons of frozen meat aboard. Of the early makers of marine refrigerating plant the most successful was Sir Alfred Seale Haslam (1844–1927) who bought up the Bell-Coleman patents. In 1889 some 2,000,000

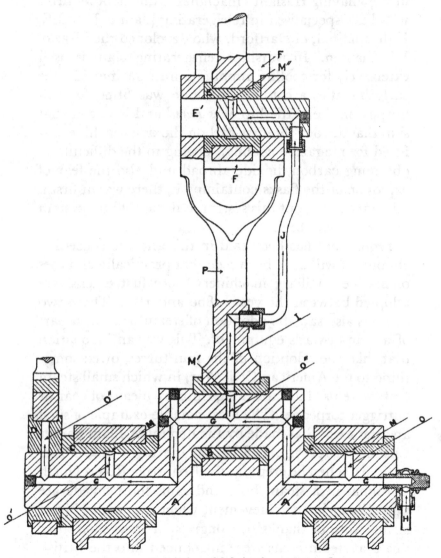

Fig. 32. Pain's forced lubrication system, 1890

carcasses of beef and mutton were brought to England in ships having Haslam's machines. Among other firms who have specialised in refrigerating plant is J. and E. Hall, Limited, of Dartford, who developed the ideas of Windhausen. In warships refrigerating plant is used extensively for cooling ammunition magazines. In the early 'nineties a cold air machine was fitted for this purpose in the battleship *Sans Pareil* and it was in that ship that in 1897 the first carbon dioxide machine was fitted for magazine cooling. Owing to the difficulty of obtaining carbon dioxide abroad and also the fear of explosion of the flasks containing it, there was at first a reluctance to adopt this system, but the cold air system has long been obsolete.

From what has been said in this and the preceding chapter, it will have been seen that practically all types of marine auxiliary machinery were first extensively adopted between the years 1860 and 1880. Those two decades also saw the adoption of steam boats as a part of a man-of-war's equipment. This was an innovation that, like the adoption of the gun turret, owed something to the American Civil War, in which small steam craft were used for attacking ships by means of spar or outrigger torpedoes. The most notable example of such an attack was that made by Lieutenant William Barker Cushing (1842–74), U.S.N., on the night of October 27, 1864, when he crept up the Roanoake River, North Carolina, in a steam boat and sank the Confederate ram *Albemarle*, an achievement which gained for him a formal vote of thanks of Congress. It was soon after this that steam boats were introduced into the British Navy, and so useful did they prove that in June 1866 the *Sailors' Home Journal and Naval Chronicle* remarked that steam launches for men-of-war were becoming as much a necessity as the engines in the ships themselves. The boat equipment of a line-of-battleship of the middle of last century included rowing and sailing cutters, pinnaces and launches, the last of which were large open boats, 42 ft. in length and 11 ft. wide. They were very stoutly built so as to be suitable for the carriage of stores, guns and ammunition. The first steam boats in the British Navy were converted launches. In the early 'sixties, Andrew Murray, Chief Engineer of Portsmouth Dockyard, fitted one of these launches with a small locomotive boiler with two single-acting vertical non-condensing engines fixed to its sides, driving twin screws. The machinery weighed about 5 tons and the speed of the boat was $6\frac{3}{4}$ knots. In 1865, Maudslays, Penns and Rennies were asked to design machinery for similar boats. Particulars and drawings of their boilers and engines are given in *Modern Marine Engineering*, by N. P. Burgh, published in 1872. The most original design was that of J. and G. Rennie, for whereas Penns and Maudslays both retained the locomotive boiler and non-condensing engines, the Rennie design included a multitubular boiler with return tubes, surface-condensing engines, and centrifugal circulating pumps driven off the engine shafts by gearing. On a trial with H.M. Steam Launch, No. 10, engined by Rennies, in Stokes Bay, June 8, 1866, a speed of 8 knots was obtained. The steam pressure was $73\frac{1}{4}$ lb. per square inch, the horse-power 32, and the weight of the machinery was 3 tons 13 cwt.

While the converted service launches had the advantages of being strong and possessing great carrying capacity, they were heavy and unhandy, and, of course, could only be carried by the larger ships. For the first improvements in the design of its steam boats, the Navy was much indebted to the Cowes shipbuilder, John Samuel White (1838–1915), and to Belliss, of Birmingham, who constructed the engines for the hulls which White built. The first steam boat supplied to the Admiralty by White was a 27-ft. cutter for H.M.S. *Sylvia*, a gunboat employed on surveying duties in the China Seas. After building this boat, White offered to construct 36-ft. steam launches with a speed of 8 knots, but with machinery weighing less than half that of the 42-ft. launches. A trial of a 36-ft. boat was successfully carried out in Stokes Bay in October 1867. The engine was a simple expansion non-condensing engine with two cylinders, $6\frac{1}{4}$ in. in diameter and 6-in. stroke. With a steam pressure of 70 lb., the revolutions were 290 per minute and the speed 8·2 knots. To reduce the noise of the exhaust steam, a "quieting tank" was fitted in the funnel. In 1878, to get more speed it was decided to adopt compound engines and forced draught. By 1880, the steam pressure had been raised to 120 lb. per square inch, and in that year trials were carried out with two 48-ft. steam pinnaces, one having a single screw and the other twin screws. Both had surface condensers. With 87 horse-power, the single-screw boat had a speed of 12·15 knots, and with 121 horse-power the twin-screw boat had a speed of 13·4 knots. To increase the manœuvring power of his boats, White then introduced the double rudder system, with the deadwood cut away. The first boat with these features was a 42-ft. steam pinnace supplied to the *Inflexible*. A further advance was made by White with what was called "Torpedo Boat, Wood, No. 5". This boat was 56 ft. long and 10 ft. wide, and the hull was constructed of two layers of mahogany, $\frac{5}{16}$-in. thick, worked diagonally and a third layer, $\frac{3}{8}$-in. thick, worked horizontally, calico painted with marine glue being used underneath the outer layer. On a trial carried out in Stokes Bay, in December 1883, the boiler pressure was 126 lb. per square inch, the revolutions 385 per minute, the indicated horse-power 142, and the speed 15·5 knots. Boats similar to this wooden torpedo-boat became known in the Navy as picket boats, and were deservedly popular. They possessed remarkably good seagoing qualities and were fitted for mounting a gun and for firing torpedoes from dropping gear. White, of course, was not the only shipbuilder to supply steam boats to the Navy, but the craft referred to illustrate the advances made during the twenty years, 1865 to 1885. By the latter year, there were some 250 steam pinnaces and launches in the Navy, 85 being carried in sea-going ships while the remainder were available for coast defence. The great majority of them were fitted for firing spar torpedoes.

## CHAPTER XVI

# TRIPLE-EXPANSION ENGINES AND WATER-TUBE BOILERS

IN the preceding chapters, the survey of the progress of the work of the engineer, as applied to ships, has been brought down to the early 'eighties of last century, by which time practically all steamships were driven by compound engines, and auxiliary steam machinery had been applied to many purposes. Just as iron had superseded wood as the material for the hulls of ships, so had steam effected a revolution in their propulsion and operation. Steamships, it is true, were still less numerous than sailing vessels, but their total tonnage was rapidly increasing, and the decline of the sailing ship had already set in. Thanks to the application of engineering science, ships were larger, stronger, safer and faster, and both for war and commerce steam was in the ascendency. But with the 'eighties a new epoch in the history of the steamship was inaugurated, for with the use of steel in the place of iron, and the adoption of the triple-expansion engine instead of the compound engine, it became possible to build ships of still greater length, size and speed, so that the ships of 1900 were as superior to those of 1880, as those of 1880 were to those of 1860.

A contemporary review of the position of the steamship in the early 'eighties was given in David Pollock's *Modern Shipbuilding and the Men Engaged in it*, 1884, while examples of marine machinery were illustrated and described in Dr W. H. Maw's handsome volume, *Recent Practice in Marine Engineering*, 1883. Both these works were published when our supremacy as builders of steamships and marine engines was unchallenged, and in one of the lists contained in Pollock's book, it is shown that of 138 steamships of over 4000 tons gross tonnage afloat, only ten had been constructed abroad—seven in France and three in the United States. Of the 138 ships, more than half had been built on the Clyde, while the majority of the remainder had come from either the Tyne, the Mersey, Barrow, or Belfast. But even when Pollock wrote, rival establishments were springing up abroad, and before the end of the century the construction of large steamships had become an international industry. Especially active were the constructors in Germany, from the yards and shops of which came some of the most remarkable steamships of the time, surpassing in size and power any constructed elsewhere, and with which she was enabled for ten years, from 1897 to 1907, to hold the blue ribbon of the Atlantic.

The progress of the steamship in the twenty years, 1882 to 1902, can be illustrated by a comparison of the Inman liner *City of Rome* with the North German liner *Kaiser Wilhelm II*. Save for the already derelict *Great Eastern*, in 1882 the *City of Rome* was the largest ship afloat. She was 560 ft. long, of 8141 tons gross tonnage, and was driven by a compound engine of 11,890 horse-power. The *Kaiser Wilhelm II*, built in 1902, was 678 ft. long, of 19,361 tons gross tonnage, and had four sets of quadruple-expansion engines driving twin screws, with a total of 43,000 horse-power. A corresponding advance was also made in the size and power of warships. In 1882, the largest cruisers had a displacement of 7000 tons and engines of 6000 horse-power. Twenty years later there were cruisers in commission of 14,000 tons displacement and 30,000 horse-power. It is with the machinery of these larger vessels that it is intended to deal in this chapter. But it should be remembered that some of the greatest achievements of the marine engineer during the period under consideration, were to be seen in the new types of vessels, torpedo boats, torpedo-boat catchers, and torpedo-boat destroyers. The development of those interesting types of vessels will always be associated with the names of Thornycroft, Yarrow and Normand, to whose work reference will be made in the succeeding chapter.

The outstanding feature in marine engineering progress during the last quarter of the nineteenth century was undoubtedly the adoption and development of triple- and quadruple-expansion engines working with steam pressures of from 150 lb. to 250 lb. per square inch. The triple-expansion engine, after the initial difficulties had been overcome, proved as superior to the compound engine as the latter had proved superior to the simple-expansion engine. It led to a further economy in fuel, a reduction in the weight of machinery and the space it occupied, and enabled engines to be built of much greater size and power. Though to-day it has been superseded in many cases by the steam turbine and the oil engine, it is still in favour for certain classes of vessels. But the adoption of the triple-expansion engine was accompanied by many other improvements, among the most notable of which was the introduction into warships of water-tube boilers in place of cylindrical boilers. Then, also, with the use of forced draught came the extended use of feed heating and of superheating, both leading to increased boiler efficiency. With the increase in the size and speed of engines, new materials were used in construction, and there were improvements in pistons, rods, shafting, valves and valve gear, bearings and methods of lubrication, and particular attention was paid to the question of balancing. With the great rise in steam pressures, all sorts of difficulties were met with and had to be overcome. New types of packing for piston and valve rods were introduced, copper steam pipes gave place to lap-welded wrought-iron pipes, and then to steel pipes, and stop valves, pipe joints and fittings generally all had to be redesigned to meet the new conditions. One very valuable improvement during the period was the introduction of solid-drawn steel boiler tubes. Then, too, lengthy experiments were carried out to determine the value of steam jacketing; research led to improved condenser design, and inventors brought about a revolution in pumping plant. The work done in experimental tanks and the introduction of progressive steam trials gave designers fresh data, and the discussions at meetings of the technical societies reflected the growth of the acknowledgment of the interdependence of abstract science and engineering.

With all the various developments throughout the twenty or twenty-five years during which the multiple-expansion engine reached its maximum size and output, before being superseded by the steam turbine, it is obviously impossible to deal. It is, therefore, proposed only to touch upon the early history of the triple-expansion engine, its application to Atlantic liners and cruisers and the contemporary adoption of water-tube boilers. A wealth of information on these and other subjects is to be found in the *Transactions* of the various technical societies, some of which contain masterly historical reviews. Among these may be mentioned the papers read by Marshall, Blechynden, McKechnie and

Maginnis before the Institution of Mechanical Engineers in 1881, 1891, 1901 and 1909, respectively, and the reviews contributed to the Institution of Naval Architects, in 1897, by Engineer Vice-Admiral Sir John Durston and J. T. Milton and, in 1911, by Engineer Vice-Admiral Sir Henry Oram. To these may be added the remarkable presidential address of Sir William White to the Institution of Civil Engineers in 1903 and the Thomas Lowe Gray Lecture on "Progress in Marine Engineering", delivered before the Institution of Mechanical Engineers, January 3, 1930, by Engineer Vice-Admiral Sir R. W. Skelton. Historical papers are also to be found in the *Transactions* of other institutions.

Like the compound engine, the triple-expansion engine had a long history before it was extensively applied to the propulsion of ships. An account of the work of the pioneers of both types is given by the distinguished French locomotive engineer Anatole Mallet (1837–1919) in his *Évolution Pratique de la Machine à Vapeur*, contributed to the memoirs of the Société des Ingénieurs Civils de France, in August and September 1910. The idea of expanding steam in three successive stages, Mallet points out, was mooted as early as 1823. Perkins suggested the same plan in 1827. One of the earliest triple-expansion engines in this country was built by Daniel Adamson (1818–90) in 1861–62. The first to apply a triple-expansion engine to a steam vessel, however, was Benjamin Normand (1830–88). In 1870, Normand built a three-stage expansion engine, which, in 1871, after the conclusion of the Franco-Prussian War, was fitted in Boat No. 30 of the Compagnie des Bateaux-Omnibus de la Seine. This engine, like some other early triple-expansion engines, had two of the cylinders placed tandem-wise and had only two cranks. For several years Normand had been designing marine compound engines, and in some degree he may be regarded as the John Elder of France, his first compound engine being fitted in the *Furet* in 1860. Subsequent to his work with Boat No. 30, Normand installed two-crank triple-expansion engines in the *Faulconeer* (1871) and *Montezuma* (1872) and to several other vessels. Normand, who was the elder brother of the eminent naval constructor Jacques-Augustin Normand (1839–1906), up till 1865 had been connected with the famous works at Havre, but after that was in business for himself. Unfortunately he lacked some of the qualities possessed by his brother, and when he died at Rouen at the early age of 58, he was in reduced circumstances. Though the importance of his work was fully recognised in France, his merits have been little appreciated in this country. Contemporary with the work of Normand in the 'seventies, was that of Ferguson, Franklin, Taylor and Kirk. In 1872, Ferguson fitted a three-crank triple-expansion engine in the launch *Mary Ann*, and in 1873 and 1874 engines of this type were fitted by Franklin in the small vessel *Sexta*, and by Kirk in the *Propontis* already mentioned. Two years later Alexander Taylor fitted a two-crank triple-expansion engine in the yacht *Isa*.

The largest of these early triple-expansion engines was that fitted in the *Propontis*, the engine of which had cylinders 23 in., 41 in. and 61 in. in diameter by 3 ft. 6 in. stroke. Steam was supplied at 150 lb. pressure from Rowan water-tube boilers. The engines themselves were successful, but serious accidents occurred with the boilers, and they were replaced by cylindrical boilers working at only 90 lb. per square inch. Under these circumstances, it was not to be expected that the *Pro-*

*pontis* would lead to other vessels being fitted with triple-expansion engines, and it remained for Kirk seven years later to demonstrate fully in the *Aberdeen* the advantage to be gained by the use of engines of this type.

Alexander Carnegie Kirk was born at Barry, Forfarshire, in 1830, and died at Glasgow, October 5, 1892. After attending Edinburgh University he was apprenticed to Robert Napier and became draughtsman to Maudslay, Sons and Field. Later on he became successively manager of the firm of John Elder and Company, Fairfield, where the *Propontis* was built, and manager of Robert Napier and Sons, the builders of the *Aberdeen*. A man of high scientific attainments, his work as a marine engineer was recognised by the bestowal upon him of the degree of LL.D. by Glasgow University. It was largely through his representations that triple-expansion engines were introduced into the Royal Navy.

The *Aberdeen* was a vessel 350 ft. long and 4000 tons deadweight capacity, built for the Aberdeen line of G. Thompson and Company, a firm which had hitherto carried on trade with Australia and other parts by sailing vessels, among which was the clipper *Thermopylae*. The *Aberdeen* was fitted with two double-ended cylindrical boilers supplying steam at 125 lb. pressure to a three-crank triple-expansion engine having cylinders 30 in., 45 in. and 70 in. in diameter and 4 ft. 6 in. stroke. The photograph of the engine as it stood in the engine shop, which forms the frontispiece of the fifth edition of Seaton's *Manual of Marine Engineering*, shows that the *Aberdeen's* engine was the prototype of thousands of triple-expansion engines constructed during the last fifty years. Few ships ever made such a successful debut or influenced future developments more than the *Aberdeen*. On trial when developing 1800 indicated horse-power the coal consumption was only 1¼ lb. per indicated horse-power per hour. Her first voyage was to Australia and back. On April 1, 1881, she left Plymouth with 4000 tons of cargo and coal, and after replenishing her bunkers at the Cape, reached Melbourne on May 14, after 42 days' steaming. Her average indicated horse-power had been about 1880 and her coal consumption about 1·7 lb. per indicated horse-power per hour. Messrs Thompson had been strongly advised by some of their friends against the experiment, but events proved that their faith in Kirk had not been misplaced, and it thus fell out that a firm of sailing-ship owners became the pioneers in the use of the triple-expansion engine for long-distance voyages. The *Aberdeen*, it may be added, had a long and successful career and the engines remained unaltered till she was scrapped soon after the Great War.

As with the screw propeller, the surface condenser, and the compound engine, the triple-expansion engine found many to condemn it, but such objections as were raised, were, to use the words of Seaton, rather the echo of old battle-cries, than the sound of new ones. Since the advent of the compound engine, large numbers of steam-driven cargo vessels of moderate size had come into existence, and it was among the owners of these that the triple-expansion engine was first regarded with favour. In a paper read before the Institution of Naval Architects on July 29, 1886, W. Parker, the Chief Engineer Surveyor of Lloyd's Register, said that since the *Aberdeen* commenced running, there had been at least 150 sets of triple-expansion engines made for the British Mercantile Marine, while about 20 sets of compound engines had been altered into triple-expansion engines. He gave a list of the ships, the dimensions of their engines, and the names of the builders, and owners. Among the last, the great mail companies figured but little. Another feature of the list was the predominance, among the builders, of the firms on the North-East Coast. The construction of cargo vessels was almost a monopoly of this district, and to the development of the triple-expansion engine many North-East Coast engineers devoted themselves. Among the prominent marine engineers of the district were F. C. Marshall (1831–1903), Alexander Taylor (1844–1906), John Tweedy (1850–1916), and Thomas Mudd (1852–98) of the Central Marine Engine Works, West Hartlepool. The marine quadruple-expansion engine had made its appearance as early as 1884, when the *County of York* was built at Barrow. Such engines, however, were for some time little used. But, in 1894, Mudd designed a five-crank quadruple-expansion engine having two low-pressure cylinders, for the *Inchmona*. The engines were only of 948 indicated horse-power, but with steam at 255 lb. pressure, superheated 60° F., the coal consumption for all purposes was as low as 1·15 lb. per indicated horse-power per hour.

The first Atlantic liners to be fitted with triple-expansion engines were the single-screw ships *Aller*, *Trave*, and *Saale*, built at Fairfield for the Norddeutscher Lloyd, just after the completion of the *Etruria* and *Umbria*. They were followed by the *Lahn* for the same line, but all four vessels were smaller and slower than the Cunard ships. Of far greater importance were the Inman liners *City of Paris* and *City of New York*, built by J. and G. Thomson at Clydebank. With these vessels, a new era in Atlantic travel opened. Both had triple-

PLATE IX

31,000 HORSE-POWER FIVE-CYLINDER
TRIPLE-EXPANSION ENGINE. *CAMPANIA*, 1893
Courtesy of *Engineering*

expansion engines and twin-screws, and a speed of over 20 knots. During the next few years, about 20 twin-screw Atlantic liners of 20 knots or more were fitted with either triple-expansion or quadruple-expansion engines. Particulars of some of the vessels are given in Table XI, the figures included in which show the gradual increase in size, power, and speed. All the vessels were fitted with cylindrical boilers, but their machinery differed considerably, the cylinders in some cases being superimposed, with the high pressure cylinders placed above either the intermediate pressure cylinders or the low pressure cylinders. In the *Campania*, the two engines of which are shown in Pl. X, for instance, each engine had five cylinders and three cranks, while in the *Deutschland*, each engine had six cylinders and four cranks, and each engine of the *Kaiser Wilhelm II*, eight cylinders and six cranks. The engines of these vessels, together with those of the *Kronprinzessin Cecile*, 1908, marked the highest development of the reciprocating engine, the birth of which will always be connected with the names of Newcomen and Watt. They were magnificent structures standing 40 ft. or 50 ft. high, and represented the acme of the art of the mechanical engineers of the nineteenth century. It is, perhaps, to be regretted that none of them has been preserved.

The application of the triple-expansion engine to warships was contemporary with their adoption for Atlantic liners. In 1884, F. C. Marshall, of Hawthorn, Leslie and Company, designed triple-expansion engines of 8000 horse-power for the Italian cruiser *Dogali*, and the following year the same type of engine was fitted in the gunboat *Rattlesnake* and specified for the battleships *Victoria* and *Sans Pareil*. The dimensions of the triple-expansion engines of successive classes of cruisers in the Royal Navy are given in Table XII. A comparison of some of the figures with those contained in Table XI show marked differences in naval and mercantile practice, especially as regards steam pressures and the length of stroke of the engines. All the liners had cylindrical boilers, but from the *Powerful* onwards, all cruisers had water-tube boilers. As to the engines, those of the *Orlando* class were three-crank horizontal engines, the last large horizontal engines fitted; those of the *Edgar* class were of the ordinary vertical three-crank type, but the engines of practically all vessels built after 1895 had engines with one high-pressure, one intermediate-pressure and two low-pressure cylinders, driving four cranks. Cylinders placed tandem-wise were out of the question for warships, while the length of stroke had to be kept as short as possible. To compensate for the short stroke the revolutions had to be increased. The revolutions per minute in the *Drake* class were 120, in the *Shannon* class 125, while the speed of revolution was still higher in some other ships. It was in the *Shannon* class and somewhat similar vessels, the last cruisers to be fitted with reciprocating engines, that forced lubrication was applied to the main bearings and crank heads. There are many ways of comparing marine machinery, but here it may be noted that in a liner of the year 1900 one ton of machinery was needed for each 6 or 7 horse-power, while in a cruiser of the same tonnage the horse-power developed per ton was 12. The machinery of the *King Alfred*, of 30,000 horse-power, for example, weighed 2500 tons, while that of the *Deutschland*, of 38,900 horse-power, weighed 5670 tons.

The introduction into the Royal Navy of triple-

PLATE X

6250 HORSE-POWER FOUR-CYLINDER TRIPLE-EXPANSION
ENGINE. H.M.S. *TRIUMPH*, 1903

Courtesy of Vickers-Armstrongs, Ltd.

TABLE XI. Twin-screw Atlantic liners with triple- or quadruple-expansion engines

| Date | Ship | Length ft. | Gross tonnage | Boiler press. lb. | Cylinders in. | Stroke ft. in. | Indicated horse-power | Speed knots |
|---|---|---|---|---|---|---|---|---|
| 1888 | City of Paris | 527 | 10,798 | 150 | 45, 71, 113 | 5　0 | 20,600 | 20 |
| 1890 | Majestic | 565 | 10,147 | 180 | 43, 68, 110 | 5　0 | 19,500 | 20 |
| 1893 | Campania | 601 | 12,950 | 165 | 37 (2), 79, 98 (2) | 5　9 | 31,000 | 22 |
| 1897 | Kaiser Wilhelm der Grosse | 627 | 14,349 | 178 | 52, 89·7, 96·4 (2) | 5　8·8 | 30,000 | 22½ |
| 1899 | Oceanic | 685 | 17,274 | 192 | 47½, 79, 93 (2) | 6　0 | 28,000 | 21 |
| 1900 | Deutschland | 662 | 16,502 | 220 | 36·6 (2), 73·6, 103·9, 106·3 (2) | 6　0·8 | 38,900 | 23¼ |
| 1902 | Kaiser Wilhelm II | 678 | 19,361 | 225 | 37·4 (4), 49·2 (4), 74·8 (4), 112·2 (4) | 5　10·86 | 43,000 | 23¼ |

TABLE XII. Cruisers with triple-expansion engines and twin screws

| Date | Class of ship | Length ft. | Displacement tons | Boiler press. lb. | Cylinders in. | Stroke ft. | Indicated horse-power | Speed knots |
|---|---|---|---|---|---|---|---|---|
| 1886 | Orlando | 300 | 5,600 | 130 | 36, 52, 78 | 3½ | 8,500 | 19 |
| 1890 | Edgar | 360 | 7,350 | 150 | 40, 59, 88 | 4¼ | 12,000 | 20 |
| 1895 | Powerful | 500 | 14,400 | 260 | 45¾, 70, 76 (2) | 4 | 25,000 | 22 |
| 1898 | Diadem | 435 | 11,000 | 300 | 34, 53½, 64 (2) | 4 | 16,500 | 20½ |
| 1900 | Drake | 500 | 14,100 | 300 | 43½, 71, 81½ (2) | 4 | 30,000 | 23 |
| 1904 | Shannon | 490 | 14,600 | 275 | 40⅝, 65½, 74⅝ (2) | 4 | 27,000 | 23 |

expansion engines supplied with steam generated in cylindrical boilers, occurred at a time when great pressure was being put on the engineering branch to reduce the weight of machinery. To meet this demand of the naval designers, among the steps taken was that of cutting down the size of boiler installations, while endeavouring to work boilers under forced draught up to an air pressure of 2 in. of water. This procedure immediately led to trouble; there were many breakdowns and not a few accidents. The principal trouble arose through the overheating of the tube ends and tube plates, which invariably led to leakage. Many efforts were made to overcome the difficulty, and a partial solution of the problem was found by fitting the tube ends

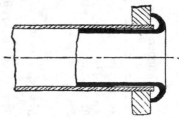

Fig. 33. Admiralty boiler ferrule

in the combustion chamber with protective ferrules. Humphrys, Tennant and Co. suggested the use of ferrules screwed into the tubes, Mr Peck, of Yarrows, ferrules leaving an air space between the ferrules and tubes, and it was partly from these suggestions that the so-called "Admiralty", "Chatham", or "Cap" ferrule was evolved (shown in Fig. 33). Made of malleable cast-iron, cap ferrules were cheap and easily renewed, and though with prolonged steaming they became clogged with scoriæ, they proved of great service and "brought more peace of mind", as one writer said, "than many a grander and more elaborate mechanical invention". But, in spite of the use of ferrules, in the end the attempt to force cylindrical boilers unduly had to be abandoned, and the maximum air pressure for forced draught was fixed at 1 in. of water and the pressure for natural draught at $\frac{1}{2}$ in. of water. The crisis which had arisen over the boiler question was partly the reason for the appointment, in 1892, of a Committee of Design, composed of both naval and civilian engineers, to report on the machinery for warships. It had been suggested that steam pressures should be lowered, but the Committee were opposed to that. On the other hand, they recommended that the amount of heating surface in ships should be increased and that experiments should be made with water-tube boilers. With the last recommendation may be said to have begun the long controversy, sometimes referred to as "the Battle of the Boilers", which, lasting more than a decade, accompanied the supersession of the cylindrical boiler and the development of the triple-expansion engine. The gunboat *Speedy* had already been fitted with Thornycroft boilers, and the gunboat *Sharpshooter* was thereupon chosen for trials with the Belleville boiler, which afterwards, for reasons sound enough at the time, was the first to be used in our larger ships. A sketch of a Belleville boiler with economiser is shown in Fig. 34. After the trials with the Belleville came others with the Niclausse, the Babcock and Wilcox, the Dürr, and the large-tube Yarrow boilers. Never before or since has the limelight of public opinion been thrown so strongly on so purely a technical problem as that of boilers for the Navy. The matter was discussed in Parliament, in the Press and elsewhere, often with little knowledge, but nearly always with great energy. Finally, however, the question was referred to a Committee composed of distinguished engineers whose labours extended over nearly four years. Their first interim report was published early in 1901*, their final

* See *Engineering*, March 15, 1901, p. 335.

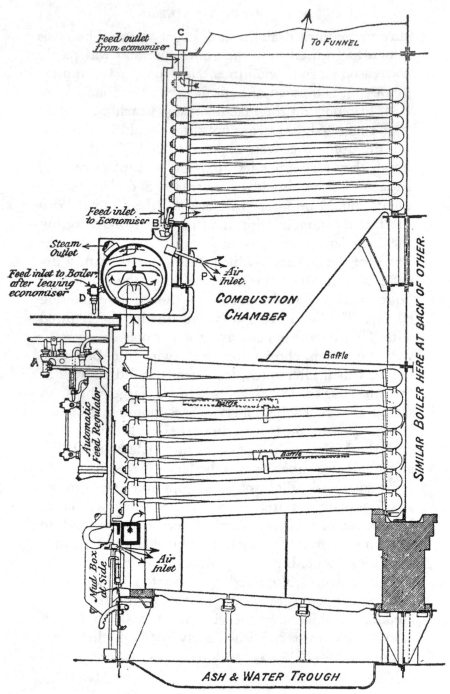

Fig. 34. Belleville boiler, 1896

report in the summer of 1904.* By the latter year water-tube boilers had been adopted by every navy of the world. The total horse-power of such boilers in the Royal Navy was about 2,500,000, in the French Navy about 1,300,000, in the Russian Navy about 800,000 and in the United States Navy about 650,000. Of the various types in warships the Belleville headed the list with a total horse-power of more than 2,000,000. These figures are evidence of the revolutionary character of the change which had taken place in the preceding ten years. Whatever the opinions held at the time, no one to-day questions the wisdom and courage of those responsible for the change.

* See *Engineering*, August 5, 1904, p. 192.

CHAPTER XVII

## MACHINERY OF TORPEDO CRAFT

TO students of naval affairs as well as to those of naval construction and marine engineering, the history of the last quarter of the nineteenth century is of especial interest, on account of the evolution during that period of the three types of torpedo craft: torpedo boats, torpedo gunboats, and torpedo-boat destroyers, introduced into all navies as the result of the adoption of the Whitehead automobile torpedo. For young naval officers, the command of these craft opened up a new and promising field of activity; and just as when the steam warship made its appearance, Commander Robinson, in his book on the steam engine, emphasised the importance of steam and urged his brother officers to study engineering, remarking that the high road to distinction and fame would be found on the paddle-box of a steamer, so half a century later, Lieutenant G. E. Armstrong, in his book *Torpedoes and Torpedo Vessels*, 1896, in speaking of the importance of a knowledge of the subject, said,

the next great naval war will bestow upon the torpedo and its users a halo of romance which will eclipse entirely that surrounding the gun and the ram. Perhaps its greatest claim to recognition on our part is the fact that it is a weapon which will nearly always be wielded by the younger officers of the Service. None know better than they that the torpedo service in war time will be the shortest ladder by which they can mount the pinnacle of fame....

For the successful attack of large ships with torpedoes it was essential that torpedo craft should possess

PLATE XI

JACQUES-AUGUSTIN NORMAND
(1839–1906)
Photo by R. Autin

SIR ALFRED FERNANDEZ YARROW,
*Bart*
(1842–1932)
Photo by Elliott and Fry, Ltd.

SIR JOHN ISAAC THORNYCROFT
(1843–1928)
Courtesy of John I. Thornycroft and Co., Ltd.

the highest possible speed, and to attain this with vessels of small displacement necessitated many departures from ordinary practice. In design, construction and equipment, these vessels were thus distinct from any other type of steam ship, and in some respects they came to represent the highest achievements of the shipbuilder and marine engineer. The weights of the hulls, the fittings and the machinery had to be reduced to the lowest possible limits, and the boilers and engines had to be capable of being worked at their maximum capacity. A comparison of the weight of marine machinery at various times shows that the total weight per indicated horse-power in early screw ships was about 600 lb., in early compound-engined ships about 400 lb., and in ships with triple-expansion engines, 200 lb. to 300 lb. In torpedo craft with reciprocating engines the weight was often as low as 50 lb. per indicated horse-power. This great reduction in weight was obtained in many ways, among which were the use of water-tube boilers of the so-called "express" type, and an increase in the speed of revolution of the engines resulting in piston speeds of as much as 1300 ft. per minute, or more. In one of the fastest destroyers, the *Express*, built by Lairds, at Birkenhead, a vessel of only 465 tons displacement, the horse-power was 9250, the engine speed 400 revolutions per minute, and the piston speed 1400 ft. per minute.

The early history of torpedo craft is inseparably connected with the work of three constructors, two Englishmen, Sir John Isaac Thornycroft (1843–1928), and Sir Alfred Fernandez Yarrow, Bart. (1842–1932), and one Frenchman, Jacques-Augustin Normand (1839–1906). Born within a year of each other, Thornycroft was the son of a sculptor, and Yarrow the son of a clerk.

Thornycroft's preparation for his life's work was obtained in the north in the shops at Jarrow and Fairfield and by contact with Rankine and Kelvin, while Yarrow served an apprenticeship with the well-known Thames firm of Miller, Ravenhill and Salkeld. When in 1866 Thornycroft at the age of 23 was able to start building steam boats at Chiswick, Yarrow and his partner had secured sufficient capital to open a small shop at Poplar and it was from the yards they established that later on came many of the finest torpedo vessels. Both introduced improvements in the construction of small craft and their machinery, both built vessels which at one time or another held the world's record for speed, and both added immensely to the reputation established for the Thames district as a centre of marine engineering by the work of the Maudslays, the Penns, Humphrys and others. Early in the present century economic forces led to the Yarrow works being removed to the Clyde and those of Thornycroft to Woolston, Hants, but the work which will be considered here was all done on the Thames.

Unlike Yarrow and Thornycroft, Normand came of a family of shipbuilders with traditions extending back to the seventeenth century. One of his ancestors had been a shipbuilder at Honfleur in 1684, while the firm known to-day as the Chantiers et Ateliers Augustin-Normand, of Havre, was founded at Honfleur in 1728. It was his father, Augustin Normand who transferred the works from Honfleur to Havre, and on his death in 1871, Jacques-Augustin Normand became the head of the firm, a position he held for 35 years. All his training had been obtained under his father, and from an early age he devoted himself to the study of the theory underlying the construction of ships. Of the many vessels constructed at Havre, an account is given in an historical booklet issued by the firm, while a review of the work of Normand himself was published by M. C. Ferrand in the *Bulletin de l'Association Technique Maritime*, of which society Normand was a leading member. His work as a constructor of torpedo boats began in 1877. He wrote many technical papers, at Havre he instituted investigations into new materials, and he was a pioneer in the practice of feed-heating, the use of feed water grease filters, and in the balancing of engines, and in 1893 he patented the well-known Normand boiler fitted in many vessels of the French, the British and other navies. Between Normand, Thornycroft and Yarrow there existed a friendly rivalry. Normand's greatness as a constructor was marked by the erection in 1911 at Havre of a striking monument to his memory, at the unveiling of which Great Britain was represented by Sir William White, S. W. Barnaby, and J. Gravell.

While it was at Chiswick, Poplar and Havre that most of the pioneering work in connection with the construction of torpedo craft was done, other men achieved success in the same direction, and afterwards many firms were engaged in the construction of torpedo vessels. Among the most prominent of these were White of Cowes, Ferdinand Schichau (1814–96) and his colleague Carl H. Ziese (1848–1917) of Elbing, and the brothers James Brown Herreshoff (1834–1930) and John Brown Herreshoff (1841–1915) of the United States, who together devised the Herreshoff coil boiler and the Herreshoff torpedo boat. Needless to say, all these constructors had colleagues to whom they owed some of their success, and among these may be mentioned John Donaldson (1841–99) and Sidney Walker

Barnaby (1855–1925), both of whom were long associated with Thornycroft.

The evolution of the torpedo boat can be traced back to the use of the spar torpedo, but the history of the torpedo boat for firing the much more effective White-head torpedo begins with the *Lightning*, constructed for the British Navy, in 1876–77, by Thornycroft. Just previously he had built the fast yachts *Miranda* and *Gitana*, and it was his work with these which caused him to be known as the pioneer of high-speed vessels. Much information about these and of the *Lightning* is contained in his paper entitled "On Torpedo-Boats and Light Yachts for High-Speed Steam Navigation", read before the Institution of Civil Engineers, on May 10, 1881. The *Lightning* was a single-screw boat, 81 ft. long by 10 ft. 10 in. in beam, and 5 ft. draught, with a displacement of 28·7 tons. She had one locomotive boiler with 525 sq. ft. of heating surface, working at 120 lb. pressure, and a compound engine with cylinders $12\frac{3}{4}$ in. and 21 in. in diameter, and 12-in. stroke, which, at 350 revolutions per minute, developed 400 horsepower, giving the boat a speed of 18·55 knots. The weight of the machinery with the water in the boilers was 10·8 tons. So satisfied were the Admiralty with the performances of the *Lightning* that they forthwith ordered twelve similar boats from various firms, and it was with one of these built by Yarrow that a speed of 21·9 knots was obtained.

Yarrow, like Thornycroft, had become well known for his small steam launches, of which between 1868 and 1875 he had built more than 300, but it was the construction of this torpedo boat which formed the starting point of his world-wide reputation. In 1878, two of his boats were regarded as the wonder of the review held by Queen Victoria at Spithead, and the following year he constructed the first sea-going torpedo boat, the *Batoum*, 100 ft. long by 12½ ft. beam, and of 500 horse-power, which attained a speed of 22 knots. With these boats the record for speed in sea-going naval vessels passed definitely to torpedo craft, by which it has ever since been held.

At first, torpedo boats in the Royal Navy were divided into two classes, first and second, the former being for harbour and coast defence, and the latter for being carried on board large men-of-war. Year by year the first-class boats of all powers showed a progressive increase in size and power. In 1885, 54 first-class boats were laid down for the Navy, 27 by Thornycroft, 22 by Yarrow, and 5 by White, the largest being 127½ ft. long and of 700 horse-power. That year Yarrow also constructed the *Kotaka* for Japan, 166 ft. long and of 1400 horse-power, while in 1886 Thornycroft laid down the *Ariete* for Spain, 147½ ft. long. This vessel, fitted with a Thornycroft water-tube boiler, attained the then extraordinary speed of 26 knots for a short time and a speed of 24·9 knots on a two hours' run. By that time all navies possessed considerable numbers of torpedo boats, France having 210, England 206, Germany 180, Italy 152, and Russia 143.

Up till 1885, the machinery of practically all torpedo boats included a locomotive boiler and a two- or three-cylinder compound engine, the steam pressure being from 120 lb. to 140 lb. per square inch. In 1879, the Herreshoffs had fitted a boat with their coil boiler (see Fig. 35), and in 1886, Thornycroft installed a water-tube boiler in one of his boats as mentioned above. The speeds quoted were not attained without considerable difficulties, and there were few boats which at some

time did not experience trouble. As a practical test of
their efficiency, in 1887 24 boats—16 Thornycroft, 4
Yarrow and 4 White—were sent on an 80-mile race
from Portland around the Ore Stone and back.
A Yarrow boat came in first, having done the passage

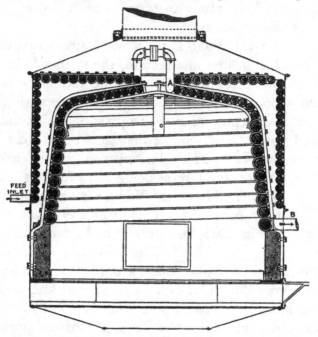

Fig. 35. Herreshoff boiler

in 5 hours 10 minutes 15 seconds. Several of them
broke down through boiler or engine trouble. Next
year out of 24 boats engaged in the manœuvres no
fewer than 16 suffered from some casualty. Locomotive
boilers as fitted in steam vessels were peculiarly liable
to fail. By some the trouble was attributed to the
shallow fire boxes, by others to the tube plates not being
protected by a bridge, while it may be that the absence
of the vibration to which a boiler of a railway locomotive is subjected when being forced at full speed

on the road, had some effect on their steaming. No
complete remedy was ever found and the locomotive
boiler was eventually superseded by various types of
water-tube boilers. The first of these to be fitted in
torpedo craft was the coil boiler designed by the
Herreshoffs in 1874, but the boilers subsequently
adopted were nearly all of the three-drum type. The
Thornycroft boiler was patented in 1885, the Yarrow

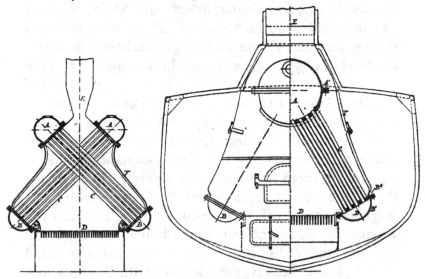

Fig. 36.  Yarrow boiler, 1889

in 1889, and the Normand in 1890. Sketches of the
two designs of boiler included in Yarrow's patent
of 1889 are given in Fig. 36. The Du Temple
boiler, used about 1890, had been originally invented
in 1876 by the French naval officer, Commandant
Felix du Temple (1823–90), for aerial navigation. The
other boilers found in torpedo craft a little later were
the Reed, 1893; the Blechynden, 1893; the Guyot,
1893, and the White-Forster, patented in 1897, the
latter being a development of the White boiler.

The existence of large torpedo-boat flotillas in all navies, with ever-increasing offensive powers, naturally called for the construction of vessels capable of countering their activities, but, though at different times a few vessels were built for this purpose, no decisive action was taken until 1886, when the Admiralty designed the first of the torpedo gunboats, or as they were often called, torpedo-boat catchers. The first of these were the *Rattlesnake, Grasshopper, Sandfly* and *Spider*, of about 525 tons and 2700 horse-power. These were followed during the next six years by 13 vessels of the *Sharpshooter* class of 735 tons and 4500 horse-power, 11 of the *Alarm* class, of 810 tons and 3700 horse-power, and 5 of the *Dryad* class, of 1070 tons and 3500 horse-power. As a whole they proved very disappointing, failing to attain sufficient speed to effect their object, and by their numerous breakdowns giving their commanding officers and engine-room staffs continual anxiety. Only one of them, the *Speedy*, constructed by Thornycroft, was originally fitted with a water-tube boiler, but later on, when the question of water-tube boilers for large vessels was under discussion, the *Sharpshooter* was fitted with Belleville boilers, the *Sheldrake* with Babcock and Wilcox boilers and the *Seagull* with Niclausse boilers. Though fitted for experimental purposes, in each case the value of the respective vessels was increased by the alteration. But after 1893 no more gunboats or catchers were laid down.

It was while these torpedo gunboats were being constructed that the French produced some of their finest torpedo boats, with speeds up to 24 knots, the inspection of which by Yarrow led to his historic conversation in 1892 with Admiral Sir John (afterwards Lord) Fisher at the Admiralty, resulting in the construction of our first torpedo-boat destroyers, a class of vessel which successfully accomplished what the gunboats had failed to do, and eventually usurped most of the functions of the torpedo boats themselves. The term destroyer was by no means new, for Sir William White had used it in a letter to *The Times* in 1884, and a writer in *The Engineer* for November 27, 1885, had again used it in referring to the *El Destructor*, then being built by J. and G. Thomson at Clydebank for the Spanish Navy. This vessel had been projected by Admiral Pezuela, the Spanish Minister of Marine. She was 192 ft. long, 25 ft. broad, and of about 385 tons displacement. She had four locomotive boilers and twin-screw triple-expansion engines, and on trial, on December 13, 1886, attained a speed of 22·6 knots when developing 3800 horse-power. She had a large radius of action, and in January 1887 created a record by steaming from Falmouth to Ferrol in 24 hours. By some she is regarded as the precursor of the torpedo-boat destroyer.

After Yarrow's report to Admiral Fisher, the Admiralty prepared a sketch design for the new type of vessel, fixing the chief features of the type, but in placing their contracts, left the responsibility for determining dimensions and guaranteeing speeds to the firms concerned, to which therefore the credit for the success of the boats belonged. The first four torpedo-boat destroyers launched were the *Havock* and *Hornet*, constructed by Yarrow, and the *Daring* and *Decoy*, built by Thornycroft, the *Havock* being launched in October 1893, and the others a few months later. The *Havock*, *Hornet* and *Daring*, each in turn, created a record for speed, the *Havock* with 26·1 knots, the *Hornet* with 27·6 knots, and the *Daring* with 28·2 knots. In general appearance they resembled large torpedo boats. The

Yarrow boats were 180 ft. long and 18½ ft. broad, with a displacement of about 220 tons, and the *Daring*, 185 ft. long and 19 ft. broad, with a displacement of about 240 tons. They were all driven by twin screws, but while the engines of the *Havock* and *Hornet* were the ordinary three-cylinder triple-expansion type, each of the engines of the *Daring* had one high pressure, one intermediate pressure and two low pressure cylinders, arranged in pairs inclined to each other, in order to reduce vibration. But the main differences in the boats were found in the stokeholds, the *Havock* having two locomotive boilers with copper fireboxes and copper tubes, the *Hornet* eight Yarrow three-drum water-tube boilers working at 180-lb. pressure, and the *Daring* three Thornycroft boilers, each having three water drums and one steam drum, working at 200-lb. pressure. This form of Thornycroft boiler became known as the "Daring" type, to distinguish it from the "Speedy" three-drum type. A sketch of the boiler is given in Fig. 37. The boilers of the *Hornet* differed from those previously used by Yarrow in having no down-comers, these being omitted as a result of some interesting experiments made at Poplar on the circulation of water and steam in tubes. There was a considerable saving in weight through fitting water-tube boilers, while, to demonstrate the capacity of a Yarrow boiler to withstand hard treatment, a trial had been carried out with one of the *Hornet's* boilers, in which steam was raised to 180 lb. in 22 minutes, and the boiler worked hard for 38 minutes, after which the fires were suddenly drawn, the fire and ash-pit doors being at the same time thrown open to the cold air. Such a test would have had serious results with most locomotive boilers, but the *Hornet's* boiler showed no signs of injury. Owing to the results obtained, all three vessels created a great deal of interest, and the following

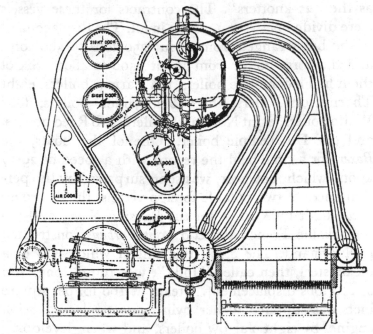

Fig. 37. Thornycroft boiler "Daring" type

particulars of the machinery of the *Daring* are worth recording for comparison with those of later vessels:

### Engines

| | | | | | |
|---|---|---|---|---|---|
| H.P. cylinder | ... | ... | ... | ... | 19 in. diam. |
| I.P. cylinder | ... | ... | ... | ... | 27 in. ,, |
| L.P. cylinders (two) | ... | ... | ... | 27 in. ,, |
| Stroke | ... | ... | ... | ... | 16 in. ,, |
| Revolutions | ... | ... | ... | ... | 387·8 ,, |
| Indicated horse-power | | ... | ... | 4409 ,, |

### Boilers

| | | | | | |
|---|---|---|---|---|---|
| Pressure | ... | ... | ... | ... | 200 lb. per sq. in. |
| Grate surface | ... | ... | ... | ... | 189 sq. ft. |
| Heating surface | | ... | ... | ... | 7890 sq. ft. |

### Condensers

| | | | | |
|---|---|---|---|---|
| Cooling surface | ... | ... | ... | 4740 sq. ft. |
| Total weight of machinery with water | | | 113·4 tons |

The *Havock*, *Hornet*, *Daring* and *Decoy*, were the first four of 42 destroyers, which, as a class, became known as the "27-knotters". The contracts for these vessels were divided between 14 firms in all. The displacement of the boats ranged from about 250 tons to 300 tons, and the horse-power from about 3500 to 4800. Six of them had locomotive boilers, ten Yarrow boilers, eight Thornycroft boilers, three Blechynden boilers, four White boilers, eight Normand boilers, two Reed boilers, and one Du Temple boiler. One of the boats, the *Boxer*, for a time held the record with a speed of 29·17 knots, which, however, was soon surpassed by the performances of two notable foreign vessels, the destroyer *Sokol*, built by Yarrow for the Russian Navy, and the remarkable French torpedo boat *Forban*, constructed by Normand. The *Sokol*, the hull of which was of a nickel steel, then called "Jolla", was 190 ft. long, and of 240 tons displacement. Steam at 160 lb. per square inch was supplied to her twin-screw triple-expansion engines by eight Yarrow boilers, and when developing 4039 indicated horse-power, she attained a speed of 30 knots, the first ship in the world to do so. The *Forban* was a much smaller vessel than the *Sokol*, being only 144 ft. 4 in. long. The stipulated speed was 29½ knots, but on trial on September 26, 1895, with a displacement of 126 tons, and her engines developing 3975 indicated horse-power, her speed was slightly over 31 knots, another world record.

The next step in the progress of torpedo craft was the construction of some 60 or 70 destroyers having displacements of from 310 tons to 370 tons, and of from 5700 to 6400 horse-power. The designed speed was 30 knots. As in the case of the 27-knotters, the 30-knotters were built by various firms, each of which fitted engines of its own design. The *Quail*, constructed by Lairds, may be taken as an example. Her displacement was 360 tons, her horse-power 6300. She was fitted with four Normand boilers working at 220 lb. pressure, her piston speed was 1101 ft. per minute, and the total weight of her machinery was 144·3 tons. On full-speed trials, the air pressure in the stokehold was 4¾ in. of water, and the coal consumption was about 2½ lb. per indicated horse-power per hour. At low speed the consumption was 1·64 lb.

The 27-knotters and the 30-knotters were all completed during the years 1893–1900, and the last years of the century saw the construction of three destroyers in which the high-speed reciprocating steam engine reached its maximum output. These vessels, the *Arab*, *Express*, and *Albatross*, were somewhat larger than the 30-knotters, as will be seen from the figures given in Table XIII, below. All destroyers had twin screws.

With these fine vessels, the development of the reciprocating steam engine in torpedo craft practically ends, for though early in this century a large number of "River" class destroyers were built of 550 tons displacement, 7000 horse-power, and 25½ knots speed, the steam turbine had already been shown to be suitable for driving high-speed craft, and after its trials in the *Viper*, *Cobra*, *Velox*, and *Eden*, it was definitely adopted for all destroyers.

A distinguished naval officer once described a torpedo boat as "a machine constructed to run a trial trip", a remark the significance of which will be appreciated by those familiar with trials of torpedo craft thirty and forty years ago. To enable such vessels fitted with reciprocating engines, to attain their designed speed, both boilers and engines had to be in the highest state of efficiency and the engine-room staff thoroughly conversant with their duties. Only the most skilled stokers could possibly maintain steam, while supervision of the engines demanded not only knowledge and experience, but presence of mind and readiness of resource in no ordinary degree. The conditions under

## Table XIII

| | Arab | Express | Albatross |
|---|---|---|---|
| Builders | J. and G. Thomson | Laird Bros. | Thornycroft |
| Length between perpendiculars, ft. | 227½ | 235 | 225½ |
| Displacement, tons | 470 | 465 | 430 |
| Boilers | 4 Normand | 4 Normand | 4 Thornycroft |
| Boiler pressure, lb. per square inch | 250 | 240 | 250 |
| Grate area, sq. ft. | 296 | 264 | 248 |
| Heating surface, sq. ft. | 16,080 | 17,020 | 16,020 |
| Cylinders, diam., in.: | | | |
| H.P. | 22 | 23 | 22 |
| I.P. | 33 | 36 | 33½ |
| 2 L.P. | 35 | 38 | 36 |
| Stroke, in. | 18 | 21 | 20 |
| Revolutions per min. | 390 | 400 | 380 |
| Indicated horse-power | 8,600 | 9,250 | 7,500 |
| Weight of machinery, tons | 207·6 | 208 | 190·3 |
| Speed, knots | 30·9 | 30·9 | 31·5 |

which trials were then carried out were in marked contrast to those obtaining to-day with oil fuel firing, steam turbines, forced lubrication and vastly improved auxiliary machinery. There was heat, noise and vibration everywhere, while in the engine room men worked in a smother of oil and water thrown off by the rapidly revolving cranks. It was, as one writer said, often a case of "pour on oil and trust in Providence". Such accidents as did occur, and they were by no means few, ranged from the bending of rods and the heating of bearings to the fracture of crankshafts and the bursting of cylinders.

Among the accidents, that to the 30-knot destroyer *Foam* on August 3, 1898, when running a trial off Malta, was one of the most memorable, owing to the action of Engineer R. W. Toman who, after one of the intermediate cylinders had burst, ordered every one out of the engine room and himself stayed below to shut off steam. "As the engines were flying round immediately after the accident", the official report said, "there was every danger of the connecting rod being driven through the bottom, but it was greatly lessened by the promptitude and pluck shown by Mr Toman in shutting off the main stop valves and so reducing the risk of the ship being sunk or seriously damaged, and the lives of all on board being probably lost." In the case of the *Bat*, another 30-knotter, it was the bottom end bolts of the low-pressure connecting rod which broke, with the result that the piston was driven out of the cylinder, projected through the deck and fell into the sea, this amazing accident occurring without injury to anyone.

The most disastrous accident was unquestionably that which occurred in the 30-knot destroyer *Bullfinch* on July 21, 1899 when doing her contractor's trials in Stokes Bay. She had already made six runs at an average speed of 29·74 knots when the starboard high pressure connecting rod broke at the fork. The cylinder fractured for two-thirds of its circumference, while a connecting rod bolt was shot through the bottom of the vessel. No fewer than eleven men were scalded to death. As the vessel was still the property of the builders a claim was made by them against the underwriters, out of which an action in the courts arose, the matter not being finally settled till about four and a half years after the accident. The case was of exceptional interest owing to the number of eminent engineers and men of science who gave evidence.*

* See *Engineering*, Nov. 21, 1902, p. 690, and Feb. 5, 1904, p. 179.

## CHAPTER XVIII

# THE INTRODUCTION OF THE PARSONS STEAM TURBINE

IN spite of the many changes in propelling machinery which were made during the nineteenth century, at certain periods engine designs tended to become more or less standardised. This was especially the case towards the end of the century, when the triple-expansion engine was supreme and without a rival. With the twentieth century marine engineering entered a period both evolutionary and revolutionary, for not only have the last thirty years seen the familiar reciprocating engine improved in design and efficiency, but they have witnessed the introduction of steam turbines for driving the propeller directly or through mechanical, hydraulic or electrical transmission gear, and also the development of various types of oil engines. But while there is thus to-day an almost bewildering variety of methods of ship propulsion, that due to the introduction of the steam turbine stands out as the most unexpected and important. Both Bramwell and Siemens in the 'eighties of last century visualised the use of internal-combustion engines aboard ship, but the advent of the marine steam turbine was foretold by none. Yet from its earliest use the turbine proved admirably adapted for driving vessels at high speeds, and turbine vessels have now held the record for speed at sea for forty years. Used in conjunction with oil-fuel burning, the steam turbine added about 10 knots to the speed of torpedo craft, and it revolutionised alike the engine-rooms of battleships and Atlantic liners; once its merits had been generally recognised its progress was phenomenal. In 1907, the total horse-power of marine steam turbines fitted was about 400,000; in 1919, the total horse-power was about 35,000,000.

The reciprocating steam engine owed its birth to the discovery in the seventeenth century of certain physical facts connected with the atmosphere. Unlike many inventions, its inception cannot be traced back to ancient times. The steam turbine, on the contrary, in a primitive form was described by Hero, whose aeolipyle is sometimes spoken of as the first reaction turbine. Many centuries later Giovanni Branca described an apparatus also utilising the energy of steam issuing from a nozzle, and this is considered the first impulse turbine. But neither of these devices was developed, and it remained for inventors in the nineteenth century to produce the first practical steam turbines. Foremost of such inventors were the Hon. Sir Charles Algernon Parsons (1854–1931), who patented his reaction turbine in 1884, and the Swedish engineer Carl Gustaf Patrik de Laval (1845–1913), who patented his impulse turbine in 1889. Since then many patents for steam turbines have been taken out, among the more important being those for impulse turbines, secured in 1896 by the distinguished French engineer, Auguste C. E. Rateau (1863–1930), and the American inventor, Charles Gordon Curtis, and that for a combined impulse-reaction turbine, secured in 1899 by Dr H. Zoelly, of Germany. Though to-day the principles of impulse and reaction turbines are utilised as circumstances demand, it was the reaction turbine of Parsons which was first extensively applied to ships.

Unlike most men of rank who devote themselves to mechanics, Parsons began life in an atmosphere calculated

PLATE XII

AUGUSTE-CAMILLE-EDMOND RATEAU
(1863–1930)
Courtesy of P. Augustin-Normand

SIR CHARLES ALGERNON PARSONS
(1854–1931)
Photo by Elliott and Fry, Ltd.

to stimulate his ingenuity. Born at 13, Connaught Place, Hyde Park, London, on June 13, 1854, he was the youngest of the six sons of the Earl of Rosse, President of the Royal Society, who was famous in the scientific world for his construction of the great 6-ft. reflecting telescope at the family seat, Birr Castle, Ireland. There Parsons passed most of his boyhood, having at one time as his tutor the astronomer Sir Robert Ball. In 1871 he entered Trinity College, Dublin, and two years later became an undergraduate of St John's College, Cambridge, taking the degree of B.A. in 1877. From Cambridge he went to the Elswick Works of Armstrong at Newcastle-on-Tyne and with this began his long association with the North of England. Leaving Elswick in 1881 he came into contact with Sir James Kitson at Leeds and two years later joined the Gateshead firm of Clarke, Chapman and Co. as junior partner. The following year he brought out his reaction steam turbine the improvement and application of which was his chief concern for the remainder of his life. Severing his connexion with Clarke, Chapman and Co. in 1889, at Heaton near Newcastle he founded the firm of C. A. Parsons and Co. for the development of steam turbines for use on land, and in 1894 the Marine Steam Turbine Company at Wallsend-on-Tyne. Although the development of the steam turbine was his chief object, he found time to pay attention to diverse scientific and mechanical problems, and in later life became known for his interest in the construction of large telescopes. His biography has been written by Mr Rollo Appleyard who, in his *Charles Parsons, His Life and Work*, 1933, says of him: "When due care had been bestowed upon a design, when due care had been exercised in construction, he put his work to the test with determination and courage.

This courage knew no bounds. It sustained his early struggles; it endured to the end." Parsons was knighted in 1911 and awarded the Order of Merit in 1927. His death took place aboard the *Duchess of Richmond* in Kingston Harbour, Jamaica, February 11, 1931. He was buried at Kirkwhelpington on March 3, a memorial service being held the same day in Westminster Abbey.

De Laval had been led to his invention through his work on cream separators; Parsons to his through his efforts to devise a high-speed engine for driving directly the newly introduced electric generators, which required a much higher speed than was then usual. His patents of 1884, comparable in importance with that of Watt of 1769, were taken out on April 23, No. 6734 being for "improvements in electric generators and in working them by fluid pressure", and No. 6735 for "improvements in rotary motors actuated by elastic fluid pressure and applicable also as pumps". The year 1884 also saw the construction of the first Parsons turbo-generator, now preserved in the Science Museum. South Kensington. This consists of a double-ended, parallel-flow reaction turbine with fixed and moving blades cut at an angle of 45°. Steam entered the casing in the middle of its length, and after passing among the blades, exhausted at the ends. The rotor revolved at 18,000 revolutions per minute, driving an armature $2\frac{5}{8}$ in. in diameter, and the machine developed about 5 kW. The design of the generator called for almost as much ingenuity as the construction of the turbine itself. Set to work in the Inventions Exhibition of 1885, this parent of all turbo-generators, was regarded at the time more as a scientific curiosity than as the forerunner of a type of engine destined to revolutionise power plants ashore and afloat. It was not long, however, before similar machines were being made in considerable numbers, and some of these found their way into ships of the Royal Navy and the Mercantile Marine.

Though for ten years after its invention the turbine was used solely for driving electric generators, Parsons had seen from the first that it could be applied to the propulsion of ships, but no steps in this direction were taken until about 1891, when it had been shown that condensing turbines could be made having a steam consumption as small as that of the best reciprocating engines. In 1894, the new development was taken up seriously, Parsons then taking out patent No. 394 for "propelling a vessel by means of a steam turbine, which turbine actuates the propeller or paddle shaft directly or through gearing". The same year saw the formation of the Marine Steam Turbine Company, the prospectus of which stated that:

The object of this company is to provide the necessary capital for efficiently and thoroughly testing the application of Mr Parsons' well-known steam turbine to the propulsion of vessels. If successful, it is believed that the new system will revolutionise the present method of utilising steam as a motive power, and also that it will enable much higher rates of speed to be attained than has hitherto been possible with the fastest vessels.

In pursuance of this object, after experiments with model boats, the company constructed the *Turbinia*. Built of steel, she was 100 ft. long, 9 ft. broad, and on a draught of 3 ft. had a displacement of $44\frac{1}{2}$ tons. She was fitted with a double-ended water-tube boiler having 22 sq. ft. of grate area and 1100 sq. ft. of heating surface and working at 210 lb. per square inch. The boiler during trials was sometimes worked under 10 in. or 12 in. of air pressure. The first set of machinery consisted of a single radial-flow turbine driving a single shaft, which, at 2400 revolutions per minute, developed 960 horse-power. The speed of the boat proving much less than had been hoped for, this radial-flow turbine was replaced by three parallel-flow turbines, one high pressure, one intermediate pressure and one low pressure, each driving a separate shaft having three screws, there being nine propellers in all. A plan of the engine room as given by Richardson is shown in Fig. 38. With steam at 157-lb. pressure at the throttle valve the

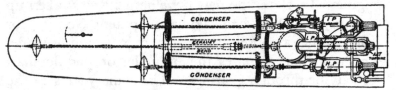

Fig. 38.  Plan of *Turbinia's* engine room

speed of the centre shaft was 2000 revolutions per minute, and that of the wing shafts 2230 revolutions per minute. On trial with these turbines the boat attained a speed of 34½ knots, or about 4 knots more than the fastest destroyers afloat. The after part of the *Turbinia* and both sets of turbines, the radial-flow and the parallel-flow, are now to be seen in the Science Museum while the forward part is at Wallsend.

The trials of the *Turbinia* were carried out off the North-East Coast, but her debut before the world was made on an occasion of pomp and circumstance, and amidst surroundings fitting in every respect for the appearance of a vessel which was to usher in a new era in ship propulsion. To mark the sixtieth anniversary of the accession of Queen Victoria, a grand Naval Review was held at Spithead, where, on June 26, 1897, was gathered together the greatest armada the world had ever seen, and it was on this occasion that the *Turbinia*, after permission had been obtained, steamed up and down the lines astonishing everyone by her performances. No fewer than 165 ships flying the White Ensign were present in the Fleet commanded by Admiral Sir Nowell Salmon, V.C., while there were also present some of the finest merchant ships, and ships of other navies. There were battleships of the *Majestic* and *Royal Sovereign* classes, the cruisers *Terrible* and *Powerful*, *Blake* and *Blenheim*, the old ironclads *Devastation* and *Thunderer*, many light cruisers, gunboats, 27-knot and 30-knot destroyers, and at one end of the lines were six sailing brigs, the last representatives of the age of masts and yards. The whole Fleet had cost over £35,000,000 and was manned by more than 38,000 officers and men. It was a magnificent spectacle and a great object lesson in the progress of naval construction and engineering. With the exception of the brigs, every vessel at anchor was fitted with reciprocating engines, but when, seventeen years later, in July, 1914, the last pre-war Review was held, all the more important ships present were driven by steam turbines in a similar manner to the little *Turbinia*.

Having thus tested and demonstrated the possibilities of the marine steam turbine, the original company transferred its business to a new company, the Parsons Marine Steam Turbine Company, Limited, of Wallsend-on-Tyne, for dealing with the new means of propulsion on commercial lines; and after negotiations, an order was obtained from the Admiralty for a turbine-driven destroyer, an event which is marked by the inclusion in the list of ships in the *Navy List* for June 1898, of the following unusual entry:

VIPER. *Twin Screw Torpedo Boat Destroyer.*
*Building at Hawthorn, Leslie & Co.'s Works for*
*Parson's Steam Turbine Co., Limited,*
*Newcastle-on-Tyne.*

The *Viper* was of the same dimensions as the 30-knot destroyers, being 210 ft. long, and of 370 tons displacement, but her turbines were estimated to develop 10,000 horse-power. Her guaranteed speed was 31 knots, but under contract conditions of coal consumption she did 33·38 knots, and on a one-hour's special trial, 36·5 knots. She was placed in commission in 1901. About the same time as the *Viper* was under construction, Sir W. G. Armstrong, Whitworth and Company laid down a slightly larger destroyer called the *Cobra*, which was fitted with a set of machinery similar to that of the *Viper*. The *Cobra*, on a three-hours' run, maintained a mean speed of 34·6 knots, and she was purchased by the Admiralty, her name appearing in the *Navy List* for June, 1900. Unfortunately, both these remarkable vessels had short careers. On August 3, 1901, during the manœuvres, the *Viper* ran ashore in the Channel Islands, and became a total wreck, while six weeks later, on September 18, 1901, the *Cobra*, while on passage from the Tyne to Portsmouth, broke in halves and foundered off the Dudgeon lightship. Out of 79 persons on board there were only twelve survivors, and among the persons lost were M. Sandison, chief engineer of Elswick Shipyard, and R. Barnard, manager of Parsons Marine Steam Turbine Company. It was afterwards suggested that the gyroscopic effect of the turbines may have contributed to the accident, but the Court Martial, which, following ordinary usage, tried the naval survivors, found that the loss of the ship was due to structural weakness, and expressed regret that she was ever purchased for His Majesty's Fleet.*

Fortunately these two disasters, which might well have had an adverse influence on the progress of the

* See *Engineering*, Oct. 18, 1901, p. 553.

turbine, had been preceded by the building of the Clyde passenger steamer *King Edward*, a vessel which demonstrated in unmistakable manner the suitability of the turbine for fast mercantile steamers. Constructed through the co-operation of Captain John Williamson (1858–1923), the founder of Turbine Steamers, Limited, the Parsons Marine Steam Turbine Company and William Denny and Brothers, Limited, Dumbarton, the *King Edward* was 250 ft. long, with a displacement of 650 tons. Her machinery consisted of one high-pressure turbine driving the centre shaft, which had two screws, and two low-pressure turbines driving the wing shafts, each fitted with one screw. On trial, in June 1901, she attained a speed of $20\frac{1}{2}$ knots; the power estimated from model experiments being about 3500 horse-power. Her appearance marked the beginning of the use of turbines in the Mercantile Marine. The success of the *King Edward* led to the construction, in 1902, of the slightly larger Clyde steamer *Queen Alexandra*, of 750 tons, 4000 horse-power and $21\frac{1}{2}$ knots; and also to the construction, in 1903, of the *Queen* for service on the Dover and Calais route. The years 1902–3 were also marked by the construction of turbine machinery for H.M.S. *Velox*, which also had two small triple expansion engines capable of being coupled to the low-pressure turbine shafts for use at low speeds, in order to increase the economy; H.M.S. *Eden*, the first vessel to be fitted with cruising turbines in addition to the main turbines; H.M.S. *Amethyst*, a triple-screw light cruiser of 10,000 horse-power; the French torpedo boat G.293; the German torpedo boat S.125, and for the German cruiser *Lübeck*.

So far there had been nothing sensational in the progress of the new type of engine and in many minds there still lingered doubts as to its future success. The position, however, was entirely altered through the work of three notable committees, two of them concerned with the construction of Atlantic liners, and the other with various classes of naval vessels. Since, in 1902, Great Britain no longer held the supremacy on the Atlantic, the Government appointed a Merchant Cruisers Committee to inquire into the question of building two large vessels, faster and with a larger radius of action than any then in existence. An agreement having been come to between the Government and the Cunard Company for the construction of two such vessels, Lord Inverclyde, chairman of the Company, in 1903 appointed a technical committee to report on the type of machinery to be used in the ships, and it was through their investigations that the famous vessels *Lusitania* and *Mauretania* were fitted with Parsons steam turbines.

Before these vessels were laid down, the Cunard Company built the turbine ship *Carmania* of 30,000 tons and 21,000 shaft horse-power, while the Allan Line, in 1904, had become the pioneers in the use of turbine-driven vessels on the Atlantic by building the *Virginian* and *Victorian* of 13,000 tons and 12,000 shaft horse-power. The step from the 21,000 horse-power of the *Carmania* to the 70,000 shaft horse-power of the *Lusitania* and *Mauretania*, which were required to have a sea-speed of 24½ knots was a very bold one, but it was fully justified by results, for, as is well known, the *Mauretania* held the Atlantic record for twenty-two years. Like all turbine vessels built at that time and for some years after, the *Mauretania* had the turbines coupled direct to the shaft and this arrangement was also seen in the notable vessels the *Imperator* (afterwards the *Berengaria*), the *Aquitania*, and the *Vaterland* (afterwards the *Leviathan*).

The arrangement of the turbines in the *Aquitania* is shown in Fig. 39. In other large ships of the Atlantic service, such as the ill-fated *Titanic*, *Laurentic* and *Britannic*, and the *Olympic*, turbines were used in combination with

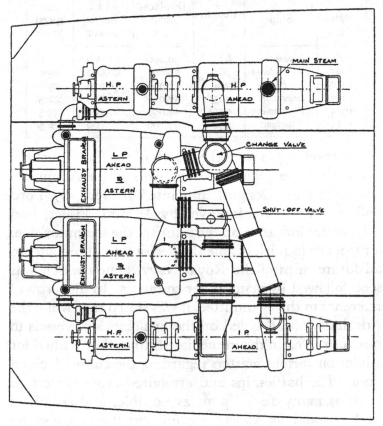

Fig. 39. Arrangement of turbines in *Aquitania*, 1914

reciprocating engines. The first important example of this arrangement was seen in the *Otaki*, built in 1908 for the New Zealand Shipping Company. Many particulars of these direct-driven turbine Atlantic liners were given in the special numbers of *Engineering* published in 1907 and 1914, dealing, respectively, with the *Maure-*

*tania* and *Aquitania*, from which many of the figures in Table XIV have been taken.

TABLE XIV.  Direct-driven turbine Atlantic liners

| Date | Ship | Length ft. | Displacement tons | Shaft horse power | Sea speed knots |
|------|------|--------|---------|---------|---------|
| 1905 | *Carmania* | 675 | 30,000 | 21,000 | 18·5 |
| 1907 | *Mauretania* | 785 | 38,000 | 68,000 | 25·0 |
| 1910 | *La France* | 714 | 27,000 | 46,370 | 23·6 |
| 1912 | *Imperator* | 905 | 57,000 | 76,250 | 22·5 |
| 1914 | *Aquitania* | 901½ | 53,000 | 60,000 | 23·5 |
| 1914 | *Vaterland* | 950 | 63,100 | 72,000 | 22·5 |

Far-reaching as were the results of the findings of the two committees referred to, the work of the Admiralty Committee on Designs, appointed in 1904 by Lord Selborne, and presided over by Admiral Fisher, had still greater influence on the future of the steam turbine, for from its inquiries came the decision to fit turbines to all future ships of the Royal Navy, a step which was soon followed by many other countries. In the terms of reference to the Committee, it was stated that while the Admiralty had decided on the five classes of vessels to be constructed in the immediate future, they wished for advice on various matters regarding the classes decided upon. The battleships and armoured cruisers were to have as many 12-in. guns as possible, and speeds of 21 knots and 25 knots, respectively; the ocean-going destroyers were to have a speed of 33 knots; the coastal destroyers 25 knots, and an experimental large ocean-going destroyer a speed of 36 knots. Such conditions made the use of steam turbines almost obligatory, and in view of the experience gained through the trials of the *Amethyst*, there was little hesitation in arriving at a decision. The first meeting of the Committee was held

in January 1905, and on October 2 the keel plates of
the historic turbine-driven battleship *Dreadnought* were
laid down at Portsmouth. Referring to the adoption of
the Parsons turbine for her propulsion, the First Lord of
the Admiralty said:

While recognising that the steam-turbine system of
propulsion has, at present, some disadvantages, yet it
was determined to adopt it because of the saving in
weight and reduction in number of working parts and
reduced liability to breakdown, its smooth working,
ease of manipulation, saving in coal consumption at
high powers, and hence boiler-room space and saving
in engine-room complement, and also because of the
increased protection provided for with this system due
to engines being lower in the ship—advantages which
more than counterbalance the disadvantages. There
was no difficulty in arriving at a decision to adopt turbine propulsion from the point of view of sea-going
speed only. The point that chiefly occupied the committee was the question of providing sufficient stopping
and turning power for purposes of quick and easy
manœuvring.

It was while these three committees were at work
that the Parsons Marine Steam Turbine Company
adopted the policy of granting licences to both British
and foreign constructors, and before long turbines of
the Parsons type were being built in France, Germany,
Austria, Italy, Russia, Spain, Belgium, Sweden, Denmark, Japan, and the United States.

The contractors for the machinery of the *Dreadnought*
were Vickers, Sons and Maxim, Limited, but the turbines were made at Wallsend-on-Tyne. The ship had
four shafts, with two high-pressure turbines and two
low-pressure turbines, with two cruising turbines
coupled to the low-pressure turbine shafts. There were
astern turbines on all four shafts. Both the ship and her

machinery were constructed with great expedition, and on October 3, 1906, 366 days after the laying of the keel plates, she began her trials. She was 490 ft. long, of 82-ft. beam, 17,000 tons displacement, and carried ten 12-in. guns in five turrets, thus being by far the most powerful ship afloat. Steam at 250-lb. pressure was supplied from 18 Babcock and Wilcox marine-type boilers. Her horse-power was 23,000 and her speed 21½

Table XV.  Direct-driven turbine battleships and battle-cruisers

| Date of design | Ship | Displacement tons | Shaft horse power | Speed, knots | Weight of machinery per s.h.p. lb. |
|---|---|---|---|---|---|
| 1905 | Dreadnought (C) | 17,900 | 23,000 | 21·28 | 184 |
| 1906 | Invincible (C) | 17,250 | 41,000 | 25·5 | 166 |
| 1909 | Lion (C) | 26,350 | 70,000 | 25·8 | 154 |
| 1912 | Malaya (O) | 27,500 | 75,000 | 25·0 | 108 |
| 1914 | Repulse (O) | 26,500 | 112,000 | 31·0 | 113 |

C = coal-fired.　　O = oil-fired.

knots, or about 3 knots more than her immediate predecessors. By fitting turbines, it is said, there was a saving of 1000 tons in displacement and £100,000 in cost. The building of the Dreadnought was immediately followed by the construction of many other direct-driven turbine battleships and battle-cruisers, of which some particulars are given in Table XV, the figures in which show the gradual increase in power and the reduction in weight of machinery in turbine-driven vessels.

The first of the smaller vessels to be constructed were the coastal destroyers and the Tribal class of oceangoing destroyers, in which not only turbines were used, but also oil fuel. The coastal boats were about the same

size and power as the earliest destroyers, but the Tribal
class were much larger than any constructed before,
their displacements being from 850 tons to 1000 tons,
and the horse-power 14,000 to 15,500. The fastest of the
class was the *Tartar* 850 tons, which attained a speed
of 35·4 knots. Between 1907 and 1915, several other
classes of direct-driven turbine destroyers were built,

TABLE XVI.  Direct-driven turbine torpedo-boat
destroyers

| Date | Class | Tons | Shaft horse power | Speed, knots | Weight of machinery per s.h.p. lb. |
|---|---|---|---|---|---|
| 1906 | Coastal | 250 300 | 4,000 | 26 | 57·7 |
| 1907 | Tribal | 850 1000 | 14,500 15,500 | 33 | 64·0 |
| 1908–9 | Beagle | 900 | 12,000 | 27 | 64·5 |
| 1909–10 | Acorn | 760 | 13,500 | 27 | 51·5 |
| 1910–11 | Goshawk | 760 | 13,500 | 27 | 48·5 |
| 1911–13 | Acasta | 935 | 24,500 | 32 | 35·3 |
| 1913–15 | M, N, O and P classes | 1108 | 25,000 | 34 | 33·6 |

all of which, except the Beagle class of 1908–9, burnt oil-
fuel, in the use of which the British Navy led the way,
as it had done in the adoption of the steam turbine. An
analysis of the results obtained with successive classes
of British destroyers was given in *Engineering* of August
1, 1919, from which some of the figures given in
Table XVI have been taken.

It is, perhaps, unnecessary here to stress the importance of the introduction into naval vessels of oil-fuel in
place of coal, which took place during the same period
as the introduction of the turbine. The advantages of
oil have often been stated. They include its higher

calorific value, the facility with which its combustion can be controlled at all powers, the absence of dust and ashes, the reduction in labour, and the ease with which it can be shipped. But it is not generally realised that the solution of the problem of burning oil in warships was only found after many years of experiment. The vaporisation of the oil before combustion, its atomisation by either steam or compressed air, all proved unsatisfactory, and at one time the United States Navy regarded the problem as insoluble. It was at the Experimental Oil Fuel Station at Haslar that the system of pressure spraying used in the Navy was evolved. In a paper on "High Speed Craft during the War", read before the Institution of Engineers and Shipbuilders in Scotland, on April 20, 1920 Mr W. W. Marriner, when referring to progress in destroyers, said:

The adoption of oil fuel was, perhaps, the next great step in the evolution of destroyers, and an enormous amount of credit is due to the Admiralty Experimental Station at Portsmouth, which brought oil-fuel burning to the perfection which is now known. One cannot help feeling that there must be some engineer officers whose names ought to be made public so that due honour might be paid them for their work in connection with one of the greatest improvements for war vessels, especially as it placed this country far ahead of the rest of the world for many years.

When many individuals are engaged in the study of the same problem it is often difficult to assign to any particular person the credit for the step which finally leads to success. This is the case with oil-fuel burning in the Navy; but it is now no secret that the evolution and development of the system introduced into the Royal Navy owed most to Chief Inspector of Machinery James Melrose (1841–1922), and Engineer-Commander

George Herbert Fryer (1868–1913), the two officers who were responsible for the experimental work done at Haslar between 1902 and 1905.

The burning of liquid fuel beneath ships' boilers was no novelty when oil fuel was adopted in the Navy. As long ago as 1834 Bourne fitted the boilers of a ship for tar-burning and in the 'sixties experiments were being made in France, Russia and the United States. At that time Admiral Jasper Henry Selwyn was the great advocate of oil for the Navy. In a paper read to the Institution of Civil Engineers in 1878 by Harrison Aydon, the author, after referring to the use of oil fuel in Russian ships on the Caspian Sea, remarked "It is reported, that all the other vessels in the Imperial Navy of Russia are likewise to be fitted for burning oil fuel". In the 'eighties the Wallsend Slipway and Engineering Company fitted the *Gretzia* for oil fuel, the Admiralty made experiments at Portsmouth and William Doxford and Sons constructed a torpedo boat, the locomotive boiler of which was fitted with thirty-one jets through which oil was injected by compressed air. That oil as a fuel was not developed earlier was largely due to the questions of supply and cost. Without attempting to do justice to the many inventors who have contributed to the success of oil-fuel burning, mention must be made of John Jonathan Kermode (1859–1931) who for forty years was associated with its progress and whose system was tried by the Admiralty in the destroyer *Surly* in 1898.

## CHAPTER XIX

## STEAM TURBINES AND
## TRANSMISSION GEAR

WHILE the application of the Parsons steam
turbine for driving propellers directly was the
outstanding achievement of the early part of
the present century, the same period saw the introduction on a less extended scale of the turbines of Rateau,
Curtis, Zoelly, Bréguet and others, and also the development of mechanical, electrical and hydraulic
systems of power transmission for coupling turbines to
propeller shafts so that the turbines could be run at a
high speed and the propellers at a low speed, thus enabling the efficiency of both turbines and propellers to
be increased. Of the three systems, mechanical reduction gearing was developed and extensively applied to
steam vessels in Great Britain, electrical transmission
was developed by the General Electric Company in the
United States, while hydraulic transmission was first
applied to ships by Dr H. Föttinger in Germany.

Among the first vessels fitted with steam turbines
other than the Parsons, were the French torpedo boat
No. 243, of 92 tons, and 1800 horse-power, and the
Russian torpedo boat *Latoshka*, ex *Caroline*, of 140 tons,
and 2000 horse-power. These vessels were built in 1904,
the former at La Seyne and the latter by Yarrow at
Poplar, and were fitted with Rateau multi-stage impulse
turbines. The inventor of these turbines, Auguste
Rateau, was born at Royan near the mouth of the
River Gironde, October 13, 1863, and died at Paris,

January 13, 1930. After passing through the École Polytechnique with distinction, he became an inspector of mines and was successively a professor at the Mining School at St Étienne and the École Supérieure des Mines in Paris. The well-known firm Société Rateau was founded by him in 1903. It was through his studies on mine ventilation that he was led to his work on centrifugal fans, turbo-blowers and steam turbines. His position in France was very similar to that of Parsons in England. At a séance solennelle held in his memory by the Société des Ingénieurs Civils de France on November 27, 1930, and presided over by the President of the Republic, Ingénieur Général Lelong, speaking of the work of Rateau for the French Navy, said: "Pendant toute sa carrière, Rateau n'a cessé d'apporter à la Marine le plus précieux concours. Son intervention a joué un rôle capital lors de la véritable révolution qui a détrôné à bord des navires la machine à vapeur alternative pour y substituer la turbine." A monument to Rateau was unveiled at the works at La Courneuve (Seine), January 17, 1931. His work on the steam turbine was begun a few years after that of Parsons. In 1892, he published his "Considérations sur les turbo machines"; in 1896 he patented his compound or multi-stage impulse turbine and at the World Exhibition in Paris in 1900, he exhibited a set of drawings of a 1000-shaft horse-power turbine for a French torpedo boat. His views on the propulsion of ships by turbines were stated in a paper entitled "The Rational Application of Turbines to the Propulsion of Warships", read at the Jubilee meeting of the Institution of Naval Architects in 1911, and in this he gave particulars of the machinery of the French triple-screw destroyer *Voltigeur*, which had a combination of reciprocating

engines and Rateau turbines. In 1910, Rateau turbines were fitted in the destroyers *Fourche*, *Faulx* and
*Magon*, built by the Ateliers et Chantiers de Bretagne,
Nantes, a firm which has since fitted Rateau turbines
in flotilla leaders with speeds exceeding 40 knots.

Like the Rateau turbine, the impulse turbine of Curtis
was also patented in 1896. As developed by the General
Electric Company, U.S.A., it was originally used for
driving electric generators. The first Curtis marine turbine was fitted in the yacht *Revolution* in 1903, and the
second in the *Creole* of the Southern Pacific Railway.
Built by the Fore River Shipbuilding Company of
Weymouth, Mass., the *Creole* was a twin-screw passenger and cargo vessel 416 ft. long. She had two turbines, each of 4000 horse-power. A much more important installation was that of the U.S.S. *Salem*, a sister
ship to the *Chester*, which had Parsons turbines, and the
*Birmingham*, which had reciprocating engines. The contract for these vessels was signed in May 1905. The
ships were 420 ft. long, and of 3750 tons displacement
when carrying 475 tons of coal and 50 tons of feed
water. The comparative trials of the three vessels provided a large amount of useful data, some of which is
given by Mr Charles de Grave Sells in Jane's *Fighting
Ships* for 1909. The designed horse-power was 16,000,
and the designed speed 24 knots. On the four hours'
full power trials, the speeds attained were: In the
*Birmingham*, 24·3 knots; the *Chester*, 26·5 knots; and the
*Salem*, 25·9 knots. In a paper entitled "A Fifty-Years'
Retrospect of Marine Engineering", Rear-Admiral C.
W. Dyson, U.S.N., said the comparative results of the
trials of the ships were as follows:

1. Economy of propulsion: At speeds up to 20 knots
the reciprocating-engine ship was superior to the vessel

with Parsons turbines, and up to 21 knots was more economical than the vessel with Curtis turbines.

2. Overload capacity of engines: The Parsons turbines exceeded moderately the Curtis turbines in this respect and very much exceeded the reciprocating engines.

3. Reliability on trials: Both types of turbines took the lead over the light high-speed reciprocating engine in this respect, the piston speed of the latter being 1200 ft. per minute.

The next vessel in the United States Navy to be fitted with Curtis turbines was the battleship *North Dakota*, of 20,000 tons displacement, and 25,000 horse-power, laid down in 1907. About this time Curtis turbines were also fitted to three or four German torpedo-boat destroyers and to the American-built Japanese battleship *Aki* and armoured cruiser *Ibuki*.

The construction of Curtis turbines for marine purposes was first taken up in this country by John Brown and Company, Limited, of Clydebank, and it was by them that the Brown-Curtis turbine, used as an alternative to the Parsons turbine in the Royal Navy, was evolved. The reasons which led to the use of Curtis turbines were (1) the desirability of acquiring experience with a turbine capable of using superheated steam; (2) the economy obtainable at low powers without the disadvantages of close-fitting parts; and (3) the simplification of the engine-room arrangements and the expectation of attaining higher propeller efficiency. After obtaining the necessary licence, the Clydebank firm constructed a complete installation for experimental purposes, the tests of which were carried out in the presence of Admiralty officials. Many improvements in design resulted from these trials and in 1909 Brown-Curtis turbines were fitted in the destroyer *Brisk*, and in 1910 in the cruiser *Bristol*, 430 ft. long, and of 4800 tons displacement. The machinery of the *Bristol* consisted of two self-contained and independent units in separate engine rooms, driving twin-screws. The designed power was 22,000 shaft horse-power and the designed speed 25 knots, but on trial, 24,227 shaft horse-power was developed, the speed of the ship being 26·84 knots.* Brown-Curtis turbines were next accepted for the cruiser *Yarmouth*, and for four destroyers building at Clydebank and from that time onward their use in the Royal Navy spread rapidly. By 1919 Brown-Curtis turbines had been used in the designs of more than 250 destroyers, 22 cruisers and seven capital ships. They were also used in the K class submarines.

Of other types of turbines used in ships, brief mention may be made of the earliest installations of those bearing the names of Zoelly, Bréguet, Melms-Pfenniger, Tosi, and Belluzzo. The Zoelly turbine was first fitted to the German destroyer G.173, of 616 tons and 10,250 horse-power, built at Kiel, and then to the German cruiser *Köln*. Zoelly turbines were installed by the French in the triple-screw destroyer *Actée*; by the American Navy in the destroyers *Mayrant* and *Warrington*, sister vessels of the *Perkins* and *Sterrett*, which had Curtis turbines; and by the Italian Navy in the destroyers *Audace* and *Animoso*. The last of these was 245 ft. long and of 655 tons displacement. With steam at 230 lb. pressure, the shaft horse-power was 20,100, and the speed 36·12 knots. Both ship and machinery were constructed by F. Orlando, at Leghorn. The Bréguet turbine, developed by the Maison Bréguet, was first applied to the French torpedo boat No. 294, of 95 tons and 2000 horse-power, and then to the French de-

* *Engineering*, Sept. 30, 1910, p. 465.

stroyer *Tirailleur*, a triple-screw vessel, which, like the *Voltigeur*, had a triple-expansion engine driving the centre shaft. The first marine installation of the Melms-Pfenniger type was that fitted in the German torpedo boat G. 166, built by Schichau at Elbing, while the same type was also used for the German cruiser *Kolberg* of 4300 tons, and 20,000 horse-power. The two turbines of Italian origin, that developed by F. Tosi and Company, of Legnano, and that invented by Professor Belluzzo of Milan, were first used in Italian torpedo craft in 1912–13. Another turbine used in the German Navy was the Bergmann, which, in 1911, was fitted to the cruisers *Magdeburg* and *Stralsund*.

When, in 1784, Watt heard of the invention by Kempelen of what was really the primitive reaction turbine, or aeolipyle, of Hero, he sent a letter full of calculations to Boulton, remarking, "So that you see the whole success of the machine depends on the possibility of prodigious velocities", and having satisfied himself that Kempelen's engine could not compete with their own, concluded by saying "In short without god makes it possible for things to move 1000 feet pr " (*i.e.* per second) it can not do much harm". The success of every turbine depends on its capacity for using steam moving at what Watt called "prodigious velocities", and this success is attained principally by high speeds of rotation. Unfortunately, the same principle does not apply to screw propellers working in water, and in all direct-driven turbine ships, the speed of the turbines was lower, and the speed of the propellers was higher than was consistent with efficiency. There was, however, no alternative to the arrangement when turbines were first used, but since then, various systems of transmission gear have been introduced with advantageous results and direct-driven turbine ships have ceased to be constructed.

It has already been stated that in his turbine patent No. 394 of 1894, Parsons included the claim for driving the propeller or paddle shaft directly or through gearing. In some early screw ships toothed-wheel gearing had been used for increasing the speed of the propeller shaft relative to the speed of the engine, but the problem now was to reduce the relative speed of the propeller shaft, and this Parsons did by the introduction of helical-toothed reduction gearing as was first used by de Laval with his land turbines. The earliest example of such gearing in a vessel was that fitted to the twin-screw launch *Charmian*, engined in 1897 by the Parsons Marine Steam Turbine Company, for Mr F. B. Atkinson. The launch was 22 ft. long, and was driven by a single parallel-flow turbine of about 10 horse-power, running at 20,000 revolutions per minute. On the turbine shaft was a helical-toothed pinion gearing into two wheels on the propeller shafts, the rate of reduction being 14 to 1. This turbine and gearing are preserved in the Science Museum. Although the advantages to be reaped by the use of reduction gearing were not lost sight of, nothing further was done in this direction until twelve years later, when the Parsons Company purchased the cargo steamer *Vespasian* for experimental purposes. Built in 1887, the *Vespasian* was 275 ft. long, and of 4350 tons displacement, and was driven by a triple-expansion engine working with steam at 150 lb. pressure. After being purchased, her engines were overhauled, trials were carried out, and then turbines were fitted in the place of the reciprocating engines. The new machinery comprised one high-pressure turbine and one low-pressure turbine side by side, each carrying a pinion with 20 double helical teeth, gearing into a wheel on the propeller shaft with 398 teeth, the ratio of the gear being 19·9 to 1. On trial and on service there was an improvement in coal consumption of 15 per cent. due to the change. For four years the *Vespasian* ran successfully, carrying coal and general cargo between the Tyne and Rotterdam. The hull was then condemned and the turbines and gearing were fitted in the *Lord Byron*.

The results of the experiments with the *Vespasian* not only showed that the Parsons turbine could be used for the propulsion of slow vessels, but indicated a means of improving the performances in high-speed ships, and, on the advocacy of Sir John Biles, the cross-channel steamers *Hantonia* and *Normannia* of the London and South Western Railway were fitted with geared turbines. These vessels, built in 1911, were 290 ft. long and of 1500 tons gross tonnage. With 6100 shaft horse-power a speed of 20·4 knots was obtained, while the steam consumption showed a considerable improvement as compared with that of a somewhat similar turbine vessel with direct-driven propellers. Another cross-channel steamer, the *Paris*, of the London, Brighton and South Coast Railway, was also fitted with geared turbines, and by the outbreak of war in August 1914, geared turbines of a total of 260,000 shaft horse-power had been constructed for mercantile vessels. The power to be transmitted in such ships was comparatively moderate, and the application of gearing to high-powered warships had to be made with caution. In the destroyers *Badger* and *Beaver*, the first naval vessels fitted with gearing, only the high-pressure and cruising turbines were geared, but in 1912 the Admiralty adopted gearing for the whole of the machinery of the destroyers

*Leonidas* and *Lucifer*, of 965 tons and 24,500 shaft horse-power, and during the war, although the construction of geared turbines for merchant vessels had to be practically abandoned, gearing was gradually adopted for all British warships, until finally single sets of gearing, as seen in H.M.S. *Hood*, were used for transmitting as much as 36,000 shaft horse-power. Of this interesting and notable step in marine engineering an account was given in a paper read by R. J. Walker before the British Association in September 1919, and in another by Engineer Commander H. B. Tostevin, R.N., read before the Institution of Naval Architects in March 1920. By September 1919, said Walker, the total horse-power of geared marine turbines completed or under construction was about 18,000,000. During the discussion of the paper Sir Eustace Tennyson d'Eyncourt said that

before gearing was used, it was impossible to accurately make the propeller to suit the turbine speed; a propulsive coefficient of about 40 per cent. was obtained. Gearing, by making the best propeller speed possible of attainment, had raised this to about 60 per cent. This was a most important improvement. It meant that with 20,000 horse-power one could now do work which without gearing would require 30,000 horse-power.

It should be added that reduction gearing as fitted in naval vessels with such success was all of the single reduction type. Double reduction gear was first fitted in 1918 in the S.S. *Somerset*.

The method of driving the propeller shaft through gearing raised anew the question of the design of thrust blocks which had been much discussed in connection with very large reciprocating engines, but which for the time had receded into the background. In direct-driven turbine ships, the thrust of the propeller was balanced partly by the reaction of the steam in the turbine, and only comparatively small ring-and-collar thrust blocks were required. With the introduction of gearing, provision had again to be made for taking the whole of the thrust, and difficulties soon arose with the old type of bearing. Fortunately, the problem had already been solved by the invention of the single-collar thrust bearing, by Mr A. G. M. Michell, F.R.S., of Melbourne. The evolution of this bearing is an example of the importance of theoretical and experimental research. The success of the Michell bearing depends on the phenomena of pressure oil-film lubrication first noticed by Beauchamp Tower (1845–1905) in his tests made during 1883–85 for the Institution of Mechanical Engineers. The subject was treated by Osborne Reynolds (1842–1912) in a paper on "The Theory of Lubrication", read before the Royal Society in 1886, but it was left for Michell to complete Reynolds's theory and show how it could be applied to journal and thrust bearings. In a paper entitled "The Lubrication of Plane Surfaces", contributed to the *Zeitschrift für Mathematik und Physik* in 1905, he dealt with the question mathematically, and then demonstrated practically that a rectangular block, pivoted at its point of resultant pressure, will automatically assume an angle to an opposing lubricated surface, depending on the speed of rubbing, viscosity of the oil and pressure. On this principle he founded his well-known thrust bearing. The introduction of this bearing into this country was mainly the work of Henry Thornton Newbigin (1864–1928), who, in a paper read before the British Association in 1916, stated that the coefficient of friction of a

Michell thrust bearing was about 0·0015, as compared with 0·03 of the multiple-collar thrust, and that such a bearing could carry, with a greater factor of safety, a load of 200 lb. to 300 lb. per square inch than the old type could carry a load of 50 lb. per square inch. The first vessel to have a Michell thrust bearing was the cross-channel steamer *Paris*, mentioned above, and the first naval vessel the destroyer *Leonidas*. It had taken many years for marine engineers to realise the possibilities of Michell's invention, but experience soon dispelled all doubts as to its value, and during the war Michell thrust blocks were fitted by the Admiralty to turbines of a total of 10,000,000 horse-power. In an action in the law courts for the extension of the patent, heard before Mr Justice Sargeant in March 1919, it was said that the saving to the nation by the use of Michell bearings during the previous four years had been of the order of £600,000, and that the construction of the *Hood* would have been impossible without it. The thrust shafts of the *Hood*, through each of which 36,000 shaft horse-power is transmitted, are 24 in. diameter, the single-thrust collars are 4 ft. 6 in. diameter and 7½ in. thick, while the bearing pads have a total area of 1176 sq. in., allowing for a thrust of about 200 lb. per square inch at maximum speed. A thrust bearing similar to that of Michell, it should be added, was simultaneously invented by Mr Kingsbury in the United States, and in that country it bears his name.

The second method adopted for indirectly driving the propeller shafts in turbine ships was that of electric transmission. Propulsion by means of electricity obtained from storage batteries had its birth in the 'eighties of last century. In 1886, the electric launch

*Volta* crossed the Channel, and the same year the submarine *Porpoise* carried out trials. Both these vessels were fitted with accumulators made by the Electric Power Storage Company, of Millwall. Since then many electric launches have been constructed, and every submarine has been fitted with batteries for use when running submerged. Very early in the present century a small vessel, running on the inland waters of Russia, was fitted with an oil engine driving a generator supplying current to motors on the propeller shafts, but electric propulsion as applied to steam vessels had its birth with the construction, in 1908, of the two twin-screw fireboats, *Joseph Medill* and *Graeme Stewart*, for the city of Chicago. Each of these vessels was fitted with two 1000-horse-power Curtis turbines, each coupled to a 250-kW direct-current generator and to a 1000-horse-power centrifugal pump. On each propeller shaft was a 250-horse-power, 220-volt electric motor. The vessels were fitted for pilot-house control and gave satisfactory results as to economy, simplicity of control and manœuvring powers. Five years later, electric transmission was tried with success in the United States twin-screw collier *Jupiter*, now the aircraft carrier *Langley*. The *Jupiter* was a sister ship to the *Cyclops*, fitted with reciprocating engines, and the *Neptune*, fitted with Parsons geared turbines. The vessels were 548 ft. long and of about 20,000 tons displacement. The *Jupiter's* main machinery included one 5500-kW Curtis turbo-generator running at 2130 revolutions per minute, and two induction motors on the propeller shafts running at 117 revolutions per minute, the reduction ratio being approximately 18 to 1. Experience with this installation led the United States to apply electric transmission to their battleships, the first of the electrically propelled vessels being the *New Mexico*, 624 ft. long and of 32,000 tons displacement. The ship was fitted with two Curtis turbo-generators running at 2070 revolutions per minute, supplying current to induction motors running at 170 revolutions per minute, there being one motor on each of the four propeller shafts. The full power developed was 31,300 horse-power, and the speed of the ship 21·3 knots. Full descriptions of the machinery of the *Jupiter* and of the battleships were given by Commander S. M. Robinson, U.S.N., in his work, *Electric Ship Propulsion*, 1922. The credit for the development of electric propulsion in America is mainly due to Mr William LeRoy Emmet, of the General Electric Company. Born in 1859, at Pelham, New York, Mr Emmet graduated at the United States Naval Academy, but abandoned a naval career for electrical engineering. When, in 1930, at the Jubilee of the American Society of Mechanical Engineers he was awarded a Society's medal, it was said that he "directed the development of the Curtis turbine by the General Electric Company; designed the machinery of the first ship driven by electric motors; was the first serious promoter of electric ship propulsion, and developed the mercury-vapour process".

The inauguration of electric propulsion of steam-ships in Europe was due to Swedish enterprise. In 1916, the Rederiaktiebolaget Svea, of Stockholm, built two sister ships, the *Mjölner* and *Mimer*. The latter was fitted with triple-expansion engines, but the *Mjölner* had two Ljungström radial-flow reaction turbines driving electric generators. The speed of the turbines was 9200 revolutions per minute and the total output 800 kW. Current at 500 volts was supplied to two induction motors placed side by side, running at 900 revolutions per minute, which drove the single propeller shaft through single-reduction gearing at 90 revolutions per minute. The same type of machinery was fitted in 1918 in the British-built S.S. *Wulsty Castle*, the first turbo-electric ship constructed in Great Britain. The machinery of the *Wulsty Castle* was also described by Commander Robinson, but after he wrote, the original machinery was replaced by Diesel engines driving the propeller shaft through a hydraulic coupling and helical gearing. The question as to the relative merits of mechanical gearing and electrical transmission for various classes of vessels has led to many discussions. No navy has yet adopted the latter for high-speed torpedo craft, and its advantages for large warships has yet to be fully demonstrated. Some interesting views on the matter are contained in Mr S. V. Goodall's lecture on "American Warship Practice", delivered before the Portsmouth Engineering Society, on January 31, 1922.*

The third system of speed reduction for turbine driven ships is that of hydraulic transmission, in which a primary water turbine wheel on the steam turbine shaft delivers water at high velocity to a secondary water turbine wheel on the propeller shaft. For the purpose of reversing, a second secondary water turbine wheel is fitted with vanes arranged in an opposite direction to those in the first. The invention of this ingenious apparatus is due to Dr H. Föttinger, of the Vulcan Works, Stettin, and the first marine installation was fitted in the tug called *Föttinger Transformator*, 96 ft. long and 76 tons displacement, with a designed speed of 12 to 13 knots. A more powerful set was that fitted in the *Königin Luise*, of the Hamburg-American Line.

* *Engineering*, Mar. 17 and 24, 1922, pp. 320, 371.

During the war, Föttinger transmitters were used in German destroyers aggregating 23,000 shaft horsepower, the maximum revolutions per minute of the steam turbine being 2500, and of the propeller shaft 520.

## CHAPTER XX

# STEAM MACHINERY
## FROM 1919 TO 1937

THE general progress with steam machinery has now been traced down to the time of the Great War, 1914–18, and it remains to review briefly the advances made during the subsequent years. During the War a vast amount of machinery was made for both naval and mercantile vessels, but the circumstances allowed little time for trying out new ideas. Ships had to be built and engined with the least possible delay. The post-War period, on the other hand, has been marked, not only by innovations in steam practice, but it has seen the rise of the large marine internal combustion engine and the construction of whole fleets of motor ships, so that whereas before the War practically all vessels, save submarines, were driven by steam, now about a fifth of the world's tonnage is driven by oil engines. The total gross tonnage of mercantile ships driven by reciprocating steam engines, steam turbines and internal combustion engines respectively, at intervals since 1919, is shown in Table XVII which has been compiled from statistics published in the *Register Book* of Lloyd's Register of Shipping, all vessels over 100 tons are included.

Leaving the history of the marine internal combustion engine to be dealt with in the next chapter, it may be said that the principal advances in steam practice have been the use of steam at higher pressures and temperatures, the wider adoption of water-tube boilers, the use of exhaust-steam turbines in combination with reciprocating engines and the extended use of oil fuel. The fuel problem has a special significance for Great Britain, which has ample coal supplies but which has to import oil. However much it is desired that British ships should use coal it is impossible to overlook the advantages of oil for many services. Every

## TABLE XVII

| Date | Steam reciprocating engines | Steam turbines | Internal combustion engines |
|---|---|---|---|
| 1919–20 | — | — | 752,606 |
| 1922–23 | 51,653,324 | 8,149,165 | 1,540,463 |
| 1926–27 | 50,040,978 | 9,137,675 | 3,493,284 |
| 1931–32 | 50,225,758 | 9,065,610 | 9,431,433 |
| 1936–37 | 42,605,474 | 9,108,812 | 12,290,599 |

Note. Prior to 1929 ships with reciprocating engines and turbines were shown under steam turbines but they are now included under steam reciprocating engines.

navy burns oil as fuel and there are now about 20,000,000 gross tons of mercantile steam vessels burning oil. Motor ships and oil-fired steam ships now comprise half the world's tonnage. With the aim of making more use of our coal quite a number of vessels have been fitted with mechanical stokers with satisfactory results. Ships have also been fitted for burning pulverised fuel and so-called "colloidal" fuel, a mixture of coal dust and oil, but neither of these systems has made much headway. With every modern improvement in steam machinery it is now possible to run steam ships on a fuel consumption of about 1 lb. of coal, or 0·6 lb. or even less of oil fuel, per horse-power per hour. As coal is far cheaper, and oil fuel slightly cheaper than the oil used in motor ships, which have an oil consumption of about 0·4 lb. per horse-power per hour, steam machinery is now better able to compete with oil engines. Further, while propelling machinery has become more efficient there has been a revolution in the design of auxiliary plant, much of which is now driven by high-speed turbines, electric motors or high-speed oil engines.

Of the 42,000,000 odd tons of shipping shown in Table XVII as fitted with steam reciprocating engines the greater part is fitted with the well-tried, reliable and robust cylindrical return-tube boiler. A few such boilers have been made for a pressure as high as 300 lb. per square inch, but they are generally constructed for pressures of from 200 to 250 lb. Incorporated with modern boilers of this type are airheaters which raise the temperature of the air for combustion to 300° F. or more, and superheaters for raising the temperature of the steam to about 650° F. The superheater tubes are sometimes placed in the smoke boxes, sometimes in the boiler tubes and in some instances in the combustion chambers. According to Mr S. Hunter, Jnr, of The North-Eastern Marine Engineering Co. Ltd., a firm which has been in the forefront in the application of superheating in merchant ships, in 1908 there were 160 ocean-going vessels using superheated steam, while in 1931 there were 4456 ships of 9,160,000 horse-power with superheaters. Feed-heating is also the general practice, the ingoing feed water being about 300° F. With airheating, superheating and feed-heating, efficiencies of over 80 per cent. have been recorded with cylindrical boilers. A sketch of a superheater as fitted in a combustion chamber by The North-Eastern Marine Engineering Co. is shown in Fig. 40.

Though since the end of last century water-tube boilers have been used in all naval vessels, shipowners as a whole have been reluctant to adopt them, and up

to about ten years ago even British-built liners retained
the cylindrical boiler. It was recognised that further
economy in fuel consumption would result from the use
of steam at higher pressures than those prevailing, and
when it was decided to raise the pressure the use of
water-tube boilers became compulsory. As with so
many other innovations, the initiative in the use of

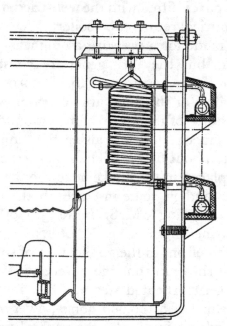

Fig. 40. Combustion chamber type superheater

steam at over 300 lb. pressure was due to Sir Charles
Parsons, at whose instigation the Clyde steamer *King
George V* was built in 1926 and fitted with Yarrow
boilers generating steam at 550 lb. pressure and 750°F.
for use in geared turbines. On trial the turbines developed 3489 shaft horse-power and the coal consumption was 1·085 lb. per horse-power per hour. Unfortunately in September 1927 a boiler tube burst and two

firemen were killed. The cause of the accident was traced to the use of impure water. In 1928 the Yarrow boilers were replaced by Babcock and Wilcox boilers, but these have since been replaced by cylindrical boilers which are considered more suitable for the intermittent service on which the vessel is employed. The original installation was in the nature of an experiment. It demonstrated fully the advantages of working with high-pressure and high-temperature steam and it forms a landmark in marine boiler practice. Four years after the *King George V* was built the Admiralty installed boilers generating steam at 500 lb. pressure and 750° F. temperature in the destroyer *Acheron* of 34,000 shaft horse-power, and on trial the oil consumption was only 0·608 lb. per horse-power per hour.

The two water-tube boilers used most extensively in mercantile vessels at present are the Yarrow and the Babcock and Wilcox. Some particulars of the Yarrow boiler as fitted in torpedo craft, for which it was originally designed, were given in Chapter XVII. Since then it has been progressively improved; from it has been evolved the "Admiralty" type of boiler used in the Royal Navy, and it is also used in large power houses. Among the most important vessels with Yarrow boilers are the *Duchess* class, and the *Empress of Britain* of the Canadian Pacific Company, the *Strathnaver* and *Strathaird* of the Peninsular and Oriental Steam Navigation Company, the Italian Atlantic liner *Conte de Savoia* and the *Queen Mary*. The pressure in the *Duchess* class is 370 lb., in the *Empress of Britain*, *Strathnaver* and *Strathaird* 425 lb. and in the *Conte de Savoia* 450 lb. and in the *Queen Mary* the pressure is 400 lb. and the temperature 700° F. A sketch of one of the 24 boilers of the *Queen Mary* is shown in Fig. 41. In these boilers the five

20-2

drums were forged from solid steel ingots. The steam drums are 21 ft. 9 in. long, 54 in. internal diameter and $2\frac{1}{32}$ in. thick; the water and superheater drums are about the same length but considerably smaller in diameter. The boiler tubes are of two sizes, 2 in. and $1\frac{3}{8}$ in.

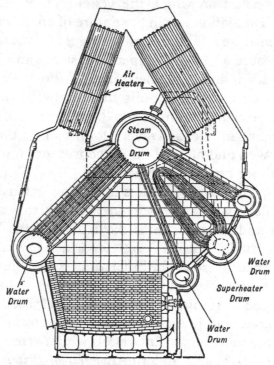

Fig. 41. Yarrow boiler of *Queen Mary*, 1936

in diameter; the superheater tubes $1\frac{1}{8}$ in. diameter. Each furnace has seven oil burners. The boilers are worked under the closed-stokehold system. The ship has also cylindrical boilers supplying steam at 250 lb. pressure for various purposes. A further step in the use of higher steam pressures is being taken by fitting Yarrow boilers working at 600 lb. pressure in the *Nieuw Amsterdam* now under construction.

The Babcock and Wilcox boiler has a very long history, and as used to-day represents the development of the boiler patented in the United States in 1856 by Stephen Wilcox (1830–93) and D. W. Stillman. Eleven years later Wilcox with his friend George Herman Babcock (1832–93) patented an improved

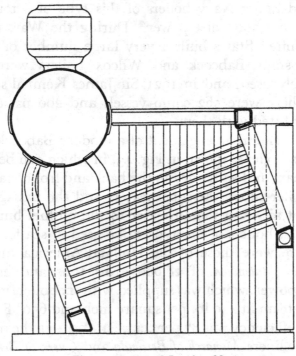

Fig. 42. Boiler of *Reverie*, 1889

form of boiler and then together founded at New York the firm of Babcock, Wilcox and Company. Their first boilers were for land purposes, but in 1891 a cross-drum sectional-header type of boiler fit for use afloat was designed, and in 1889 one of these was fitted in the yacht *Reverie*. A sketch of this boiler is given in Fig. 42. The pioneers of the use of these boilers in Great Britain were Thomas Wilson and Sons of Hull who after trying one boiler in the *Nero* in 1891, subsequently fitted them in the *Hero* and other vessels. The United States Navy first used Babcock and Wilcox boilers in the gunboats *Marietta* and *Annapolis* in 1896 and the Admiralty about the same time used them in the gunboat *Sheldrake*. By 1903 there were in use in the Royal Navy and the United States Navy boilers of this type of a total of about 500,000 horse-power. During the War, when the United States built a very large number of merchant ships, Babcock and Wilcox boilers were extensively used, and in 1921 Sir James Kemnal stated that there were 562 cargo vessels and 200 passenger vessels fitted with them.

The general arrangement of a modern Babcock and Wilcox boiler is shown in Fig. 43, in which can be seen the steam and water drum, the back and front headers, the boiler tubes, the superheaters and the passage for the hot air from the airheater to the oil-fuel burners. The first boilers of this type to be used at sea for high pressure were those of the Holland-America turbine steamer *Statendam* of 28,000 gross tons and 22,000 horse-power, which was supplied with six boilers generating steam at 430 lb. per square inch and 650° F. The boiler efficiency was 87 per cent. In 1931–32 the turbo-electric liners *Monarch of Bermuda* and *Queen of Bermuda* were fitted with boilers for 400 lb. pressure, and in 1934 the P. and O. liner *Strathmore* and the Orient liner *Orion* were fitted with boilers for 450 lb. pressure; and since then their sister ships *Stratheden*, *Strathallan* and *Orcades* also. Altogether there are now over 1300 mercantile vessels of various classes with the Babcock and Wilcox sectional-header boiler.

Though the Babcock and Wilcox, the Yarrow and other three-drum boilers have secured a recognised

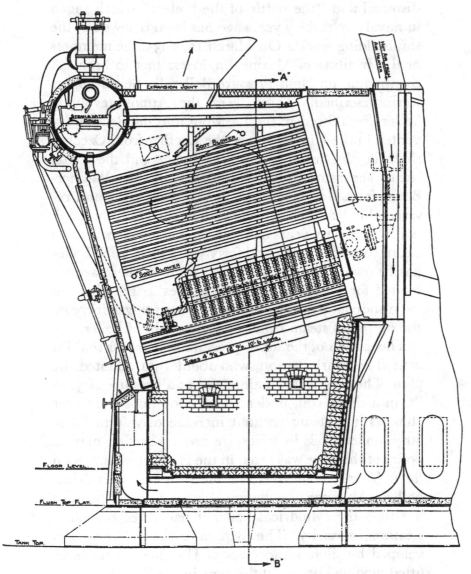

Fig. 43. Babcock and Wilcox sectional-header boiler

place in mercantile fleets, their supremacy is not un-disputed and "the battle of the boilers" which raged in naval circles forty years ago has been renewed in the shipbuilding world. On March 12, 1935, the members of the Institute of Marine Engineers met to hear "A Symposium on High Pressure Boilers". Seven boilers were described. These were the Atmos revolving boiler, from Sweden, the La Mont and the Benson, in-vented in the United States, the Loeffler from Czecho-slovakia, the Sulzer mono-tube boiler and the Velox steam generator from Switzerland and the Wagner-Bauer from Germany. In some of these are found re-vived old ideas such as were seen in the Herreshoff coil boiler of seventy years ago and the flash boilers used in steam road cars thirty and forty years ago. Each boiler has its own special features and some of them are in-tended for so-called "super-pressures". But they all have one common object—that of attempting to increase the output of steam from a given weight of boiler.

The earliest of the super-pressure boilers was that in-vented by Mark Benson, who about 1921 adopted the plan of heating water maintained at a pressure of 3200 lb. in a single coil, under which conditions the water changes into steam without increase of volume. For large boilers coils in series are used. The first marine boiler of this type was fitted in the Hamburg-American liner *Uckermark* in 1931, in place of one cylindrical boiler. Although the Benson boiler weighed 60 tons less than the cylindrical boiler it had three times the steaming capacity. The Benson boiler is being de-veloped by Blohm and Voss of Hamburg, who have fitted modified boilers of this type in the Norddeutscher liner *Potsdam*, the steam pressure being 1325 lb. and the temperature 878° F.

In the boiler invented by Dr Loeffler the heat from the furnace is transmitted to tubes through which superheated steam is circulated by a pump. Some of the superheated steam passes to the turbines and some is returned to a water drum, not exposed to the hot gases, in which saturated steam is generated. A Loeffler boiler for 1830 lb. pressure, together with an additional high-pressure turbine, has been fitted in the Italian liner *Conte Rosso*, in place of a cylindrical boiler, and the speed of the ship considerably increased thereby.

The Sulzer mono-tube boiler, as its name implies, consists of a single long coil of tubing into one end of which the water is pumped and from the other end of which the steam issues. After experiments ashore with a boiler containing a mile and a half of 2 in. tubing, in 1935 a Sulzer boiler was installed in the Rotterdam Lloyd steamer *Kertosono*, the boiler having five times the steaming capacity of the one it displaced. The same method of circulating water through coils by means of a pump is used in the La Mont boiler. Of the others, the Atmos boiler is remarkable for its revolving cage of firebars and boiler tubes in which the fuel is burnt, while in the Velox boiler combustion takes place under a pressure of about 25 lb. per square inch. The Wagner-Bauer boiler more closely resembles the three-drum type than either of the others, but it has been designed for a high rate of heat transmission. The Wagner boilers in the new German liners *Gneisenau* and *Scharnhorst* generate steam at 737 lb. pressure. Another new boiler, of British origin, is that invented by Mr J. Johnson. One of his boilers was installed in the *Empress of Britain* in 1931. Three years later Johnson boilers, to the design shown in Fig. 44, for generating steam at 385 lb. pressure and 700° F. were fitted in the *Alcantara* and the *Asturias*

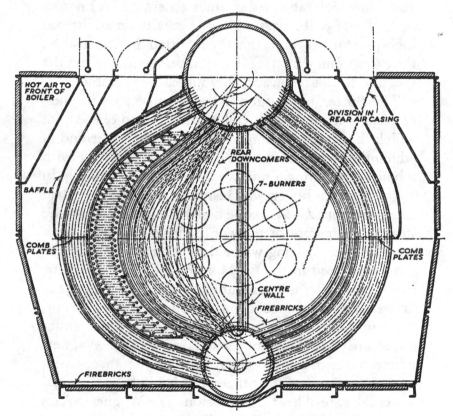

Fig. 44. Johnson boiler. (Courtesy of *Engineering*.)

when the 20,000 horse-power oil engines were taken out and replaced by geared turbines in order to increase their speed. Boilers of the same type—now known as Babcock-Johnson boilers—are being installed in the liners *Arundel Castle*, *Windsor Castle* and *Andes*.

Diverse as are the types of marine boilers, no less varied are the designs of propelling machinery, for threatened on the one hand by the steam turbine and on the other by the oil engine, the steam reciprocating engine, so long the faithful servant of shipping, has been redesigned and its efficiency considerably increased. For large vessels, steam reciprocating engines are no longer constructed, but for medium-sized and smaller vessels of moderate speed there are several types of engines to choose from. Among these are the "Quadropod" drop-valve engine built by the Central Marine Engine Works, West Hartlepool; the Semi-Uniflow engine of Alexander Stephens and Sons Limited, Glasgow; the Christiensen engine developed by Christiensen and Meyer of Germany; the Lentz engine as built by John Dickinson and Company, and the "Reheater" triple-expansion engine of The North-Eastern Marine Engineering Company Limited. The "Quadropod" engine is a quadruple-expansion engine with a patented system of steam distribution and with cam-actuated double-beat drop valves to each cylinder instead of the older slide or piston valves; the engine of Alexander Stephens and Sons is a triple-expansion engine having the low-pressure cylinder designed on the uniflow principle; the Christiensen engine is a four-cylinder double-compound engine with uniflow low-pressure cylinders and the Lentz engine is also a double-compound engine but with separate inlet and exhaust valves of the double-beat type to each end of the cylinders. In the triple-expansion engine of The North-Eastern Marine Engineering Company there is a re-heater for the steam as it passes from the high-pressure cylinder to the intermediate-pressure cylinder; the re-heater being heated by the steam at 750° F. as it passes from the boiler on its way to the high-pressure cylinder. The reheating results in a diminution of the wetness of the steam in the intermediate cylinder.

Of perhaps greater importance than these innovations is the adoption of exhaust-steam turbines in combination with reciprocating engines whereby a greater power is obtained from a given amount of steam. The system was originated by Sir Charles Parsons, the first ship having an exhaust steam turbine being the *Otaki* mentioned on p. 281. In this and other vessels such as the *Olympic* the exhaust-steam turbine drove its own particular shaft. Exhaust-steam turbines are now either coupled to the shaft of the reciprocating engine or used in other ways. The system best known is that due to the collaboration of Dr Gustav Bauer and Dr Hans Wach, and known as the Bauer-Wach system. In this system the steam from the low-pressure cylinder passes to an exhaust turbine which is coupled to the propeller shaft through double-reduction gearing and a Vulcan hydraulic coupling. First fitted in the steam trawler *Sirius*, now the *Hanseat*, in 1926, the success of the system soon led to its adoption in other vessels, in many instances an exhaust turbine being added to existing machinery. There are now some 500 vessels with the Bauer-Wach system, the total horse-power being about one and a half million. The largest vessel fitted with the Bauer-Wach system is the P. and O. twin-screw steam ship *Maloja*, 20,914 tons and 17,500 horse-power. It is claimed that the use of an exhaust turbine in the manner described above leads to an economy in fuel consumption of as much as 20 per cent.

The exhaust-steam turbine system developed by Brown, Boveri and Company is somewhat similar to the Bauer-Wach system, but others differ considerably. In that introduced by Metropolitan-Vickers Electrical Company, Limited, the exhaust turbine drives an electric generator which supplies current to an electric motor on the propeller shaft; in that brought out by the Swedish firm Aktiebolaget Gotaverken and known as the "Gotaverken" system the turbine drives a turbo-compressor which takes steam from the exhaust of the high-pressure cylinder and after raising its pressure and temperature returns it to the intermediate-pressure receiver. In the "Lindholmen" system of the Aktiebolaget Lindholmen Motala, Gothenburg, the exhaust turbine drives an electric generator the current from which is used in an electric heater through which the steam passes on its way from the high-pressure to the intermediate-pressure cylinder. In yet another system introduced by the Danish Elsinore shipyard the exhaust turbine is coupled to the propeller shaft by combined single reduction gear and a chain drive. The latest of these combined reciprocating engine and turbine plants is that invented by Mr W. A. White, who has called his plant the "New Economy" Engine. In this a high-speed double-compound reciprocating engine and an exhaust turbine both drive the propeller shaft through reduction gearing. The first example of the White system was fitted in the *Adderstone* ex *Boswell* in 1934, which had previously been driven by turbines with double-reduction gearing. The saving in fuel effected by the conversion was stated to be 40 per cent.

Whatever the future holds for the steam reciprocating

engine, or the internal-combustion engine, it appears likely that for high-speed vessels such as battleships, cruisers, destroyers and Atlantic liners the steam turbine will continue to hold the foremost place. It is true, as can be seen from Table XVII, that during the past ten years the total tonnage of turbine-driven merchant vessels has remained stationary; but that total includes many of the most important vessels in the world, and all the largest ships on the Atlantic. Turbine-driven vessels can be divided into four groups. First a small but steadily diminishing number of vessels fitted with direct drive as seen in the *Berengaria*, *Aquitania*, *Paris* and *Ile de France*; secondly a considerable number of vessels with double-reduction gearing; thirdly a much larger group of vessels with single-reduction gearing, and lastly a small but slowly increasing number of vessels with electric transmission between turbine and propeller.

Of the direct-driven turbine ships particulars were given in Chap. XVIII, and on these it is unnecessary to dwell. The first vessel with double-reduction gearing was the *Somerset*, of 4500 horse-power, built in 1918. After the War this type of gearing was fairly widely used and in his Howard lectures to the Royal Society of Arts in 1923 Mr S. S. Cook said that there had been over 200 British-built mercantile vessels fitted with double-reduction geared turbines. In some of these vessels the gearing gave a certain amount of trouble and there was a controversy as to whether the cause of these failures was due to design, material or construction. Some of the failures were traced to errors in gear cutting, but with improvements in the gear-cutting machines introduced by Sir Charles Parsons, much of the trouble disappeared. In some designs of gearing the

PLATE XIII

20,000 HORSE-POWER GEARED TURBINES
H.M.S. *DIOMEDE*, 1917

Courtesy of Vickers-Armstrongs, Ltd.

bearings of the pinion shafts were carried in what was called a floating frame which allowed for small errors in alignment, but such an arrangement has been found to be unnecessary. The pioneers of the floating frame were Rear-Admiral George W. Melville (1841–1912), who had been Engineer-in-Chief of the United States Navy, and John Henry Macalpine (1859–1927).

The wider adoption of single-reduction gearing in place of double-reduction gearing for merchant ships dates from 1923, shipowners no doubt being influenced in their decision to use single-reduction gearing by the consistently good results obtained with this type in H.M. ships. An example of a set of turbines with single-reduction gearing as fitted in warships is shown in Pl. XIII, on which is given two photographs of one set of the turbines of H.M. cruiser *Diomede*, 4850 tons, 40,000 horse-power, built in 1917. The upper picture shows the turbines and gearing with the covers removed and the lower with the plant as ready for installing in the ship. Though there are various ways of disposing the turbines and gearing relative to each other, it is with this type of machinery the *Bremen, Europa, Empress of Britain, Rex, Conte de Savoia* and *Queen Mary* are driven. A few particulars of four of these vessels are given in Table XVIII, in which are also included particulars of the *Ile de France*, the last of the great liners fitted with direct-driving turbines, and of the *Normandie* which has electric transmission gear.

It is outside the scope of this volume to deal at length with these vessels, their machinery and their voyages, but a few details relating to the *Queen Mary* cannot but be of interest. The propelling machinery consists of four independent sets of turbines, each set comprising four ahead turbines and two astern turbines. These have been designed to take steam at a pressure of 352 lb. per square inch and to exhaust at a pressure of 0·5 lb. per square inch absolute. Impulse blading is used for the first stage of the high-pressure turbines and the astern turbines; the remainder is of the reaction type. The

TABLE XVIII

| Ship | Ile de France | Bremen | Empress of Britain | Conte de Savoia | Queen Mary | Nor-mandie |
|---|---|---|---|---|---|---|
| Year of completion | 1927 | 1929 | 1931 | 1932 | 1936 | 1935 |
| Where built | St Nazaire | Bremen | Clyde | Trieste | Clyde | St Nazaire |
| Length between perpendiculars (ft. in.) | 757 10½ | 887 11 | 730 0 | 788 0 | 965 0 | 962 0 |
| Draught (ft. in.) | 31 0 | 33 10½ | 32 0 | 30 6 | 36 6 | 36 7½ |
| Gross tonnage | 43,153 | 51,656 | 42,348 | 48,502 | 80,774 | 79,000 |
| Displacement (tons) | 40,350 | 54,750 | 41,000 | 40,000 | 71,800 | 67,000 |
| Service horse-power | 52,000 | 100,000 | 62,500 | 90,000 | 158,000 | 160,000 |
| No. of boilers | 32 | 20 | 9 | 10 | 24 | 29 |
| Boiler pressure | 230 | 340 | 425 | 450 | 400 | 400 |
| Service speed (knots) | 23½ | 27 | 24 | 26½ | 28½ | 28½ |

impulse blading is of stainless steel machined to shape; the reaction blading of low-carbon stainless iron rolled to shape. The four main condensers are of the Weir regenerative type. Each is about 26½ ft. high and 18 ft. wide and contains 13,780 solid-drawn 70/30 cupro-nickel tubes ¾ in. outside diameter and 15 ft. 6 in. long. There are eight circulating pumps driven by 285-horse-power electric motors. The combined capacity of these pumps is nearly 900 tons per minute. In the closed-feed system are eight extraction pumps, eight sets of steam-jet air ejectors, twelve feed-heaters and eight turbo-feed pumps. The temperature of the feed water as it enters the boilers is 370° F. Besides four large evaporators each having an output of 100 tons a day the ship has a large water-softening plant. The auxiliary machinery throughout this ship is electrically driven, the current being supplied by seven 1300 kW turbo-generators and in the ship are no fewer than 578 electric motors of an aggregate of about 18,000 horse-power.

In the nine million tons of mercantile vessels shown in Table XVII as driven by steam turbines are included about half-a-million tons of ships which have electric transmission gear instead of mechanical reduction gearing. Although as already stated electric propulsion was adopted in the United States Navy for her battleships built about the time of the War, and for the battle-cruisers, now the aircraft carriers, *Lexington* and *Saratoga*, vessels of 33,000 tons displacement, 180,000 horse-power and a speed of 33 knots, in mercantile fleets the turbo-electric system made little headway. During the last ten years, however, the system has been adopted for several notable vessels, and whereas Lloyd's *Register Book* of 1925–26 showed only 24 ships of a total tonnage of 94,853 as driven by steam turbines and electric motors, the corresponding figure in the *Register Book* for 1935–36 are 42 ships with a total tonnage of 482,563 tons. This system having been used for the great French liner *Normandie*, it will no doubt receive still further attention. A review of the progress made up to 1932 was given by Dr C. C. Garrard in a paper on "The Electric Propulsion of Ships", read to the British Association in that year, while two years later the subject was treated from a different point of view in a paper by Mr C. W. Saunders read to the Institute of Marine Engineers.

With the increased use of electricity for driving auxiliary machinery in ships, marine engineers have become more familiar with electrical engineering practice, and it is admitted that in the matter of reliability, flexibility and economy the system leaves little to be desired. Among the advantages of electric propulsion Dr Garrard included (1) increased overall efficiency, (2) higher speeds in heavy weather, (3) flexibility in the location of machinery, (4) reduced vibration and noise, (5) increased safety by allowing better subdivision, and (6) the avoidance of reversing turbines.

The first large passenger vessels with turbo-electric propulsion were the *California, Virginia* and *Pennsylvania*, built in 1928–29 for the American Line Steamship Corporation. These vessels are about 20,000 gross tons, horse-power with a speed of 18 knots. While the *California* was being built the Peninsular and Oriental Steam Navigation Company adopted the system for the *Viceroy of India*, which had been originally designed to be fitted with geared turbines. Built on the Clyde by Alexander Stephens and Sons, Limited, the *Viceroy of India* is 585 ft. long and of 19,648 gross tons. She has six Yarrow boilers supplying steam to two 9000 kW turbo-alternators delivering current at 2720 volts to two 8500-horse-power electric motors on the propeller shafts. With full power the ship can steam 19 knots, but on certain passages only one turbo-alternator is used, giving a speed of 16½ knots.

Other notable vessels with turbo-electric drive are the *Strathaird* and *Strathnaver*, the Furness Withy liners *Queen of Bermuda* and *Monarch of Bermuda* which run between New York and Bermuda, the *President Hoover* and *President Coolidge* of the Dollar Steamship Line and the *Potsdam* and *Scharnhorst* of the Norddeutscher Lloyd.

The fullest development of the turbo-electric system is seen in the *Normandie*. This vessel has 29 Penhoët three-drum water-tube boilers supplying steam at 400 lb. pressure and 680° F. temperature to four 34,250 kW turbo-generators, delivering three-phase current at 5500–6000 volts to four propulsion motors capable of developing 160,000 horse-power. The blading of the turbines is made from A.T.V. steel (acier turbine à vapeur), a steel developed by the joint research of Hadfields Limited of Sheffield and the Commentry-Fourchambault Co. of France. The turbo-generators revolve at 2430 revolutions per minute and the pro-pulsion motors at 358 revolutions per minute. If de-sired, the ship can steam at reduced speed with two of the generators shut down. For supplying current to the auxiliary machinery she has six 2200 kW geared turbo-generators, generating current at 220 volts. In magni-tude the machinery of this ship, it will be seen, rivals that found in a large power station ashore.

## CHAPTER XXI

# THE MARINE INTERNAL COMBUSTION ENGINE

WHILE Parsons was engaged with the invention and application of the steam turbine, other inventors were devoting themselves to the improvement of internal combustion engines, and remarkable as have been the results of the adoption of the marine steam turbine, scarcely less notable have been the changes in marine propulsion brought about by the development of spirit and oil engines. With petrol engines, small racing craft are driven at speeds equalling the fastest railway trains; petrol, paraffin and heavy-oil engines have practically superseded boilers and steam engines for small craft in use on rivers and in harbours, while the large heavy-oil engine has successfully challenged both steam engines and steam turbines as prime movers for all but the fastest classes of ships.

Like that of the reciprocating steam engine, the history of the internal combustion engine goes back to the seventeenth century, but the real development of such engines did not begin until the latter half of the nineteenth century, when the manufacture and distribution of coal gas for lighting purposes had become widespread. The first experiment with any form of internal combustion engine for driving a boat, however, dates back to 1827. In 1823, Samuel Brown took out a patent for a gas vacuum engine. A company was formed to exploit the invention, in 1826 an engine was fitted in a boat 27 ft. long, and on January 1 and 31,

1827, trials were made on the Thames. Descriptions of the engine had appeared in the *Mechanics' Magazine*, and in a statement contained in that journal in 1827, Brown said, "On the 31st January a second experiment was made on the river before the Lords of the Admiralty, and a number of scientific men and the result was such as to decide their minds in favour of its eligibility." But the optimism of Brown was not shared by his supporters, the company was wound up and little more was heard of the gas vacuum engine or its inventor. After the fruitful work of Lenoir, Langen, Otto, Clerk and others about half a century later in connection with the gas engine, and the invention of the gas producer, the matter assumed another aspect and opinions were expressed favourable to the use of gas engines in ships. In 1882 Sir William Siemens, when President of the British Association, said:

Before many years we shall find in our factories and on board our ships, engines with a fuel consumption not exceeding 1 lb. of coal per effective horse-power per hour, in which the gas producer takes the place of the somewhat complex and dangerous steam boiler.

The gas producer appeared to open up many possibilities and to some engineers it seemed that the steam engine would have to give way to its younger rival. Gas engines working with producer gas indeed were later on tried many times, among the most notable experiments being those made by Beardmore and Co., Ltd., in the old gunboat *Rattler* and by A. C. Holzapfel in the *Holzapfel I*. The *Rattler* in 1907 was fitted with a gas producer and a 500 horse-power five-cylinder engine working at 120 revolutions per minute. The results were satisfactory, but writing of this experiment in his review of marine engineering progress in Jane's *Fighting Ships* in 1909 Mr C. de Grave Sells said:

The undoubted advantages of the system over steam machinery for such an installation will probably lead to its being adopted for small vessels for harbour use and vessels having a regular service from a fixed base, but there seems no probability of its extensively displacing steam machinery under ordinary conditions.

If coal were the only fuel available no doubt more would have come of the experiments with marine gas engines, but the discovery of petroleum in the United States in 1859, and the subsequent development of the oil industry in various countries made available a much more convenient fuel for use in internal combustion engines. The first to construct engines using petroleum appears to have been George B. Brayton (1830–92) of Philadelphia, who took out a patent in 1874. Nine years later, in Germany, J. Spiel patented a petroleum spirit engine, and the same year Gottlieb Daimler made the first high-speed spirit engine, the forerunner of the engines used in motor cars, aeroplanes and fast motor boats. In 1885 Messrs Priestman Brothers of Hull began the development of the Etève engine, which became the first successful engine to use commercial lamp oils.

In all these engines the charge of air and vaporised spirit or oil was fired by an electric spark, a flame or a heated tube. In the modern heavy-oil engine neither of these devices is used, ignition of the charge being caused by the heat due to the compression of the air in the cylinder. They are thus called compression-ignition engines, and they all spring from the pioneering work of Stuart and Diesel in the 'nineties of last century. Their inventions proved the turning-point in the history of the oil engine and though the many forms of oil

PLATE XIV

HERBERT AKROYD STUART
(1864–1927)
Courtesy of Blackie and Son, Ltd.

RUDOLPH DIESEL
(1858–1913)
Courtesy of Deutsches Museum

engines in use to-day have only been brought to their present state of perfection through the work of a great many individuals and by the expenditure of perhaps millions of pounds, by Governments, firms and institutions, the names of Stuart and Diesel will ever remain associated with oil-engine history.

Herbert Akroyd Stuart was born in Yorkshire in 1864 and died at Claremont, West Australia, February 19, 1927. He is buried in the Akroyd Cemetery, Halifax, Yorkshire. After attending Newbury Grammar School and passing through Finsbury Technical College he entered his father's engineering works at Fenny Stratford, Buckinghamshire. In 1886 at the age of twenty-two he took out the first of several patents in connection with oil engines, the most important of which were No. 7146 of May 1890 and No. 15,994 of October 1890. In the first of these he described an engine working on the "Akroyd" cycle and provided with an extension of the cylinder known as the vaporiser, which on starting the engine had to be heated by a lamp. It worked on the four-stroke cycle introduced by Otto for gas engines. Air was drawn in on the suction stroke and compressed to about 30–35 lb. per square inch on the return stroke. At the end of the compression stroke the oil was injected into the vaporiser by a pump, combustion took place, the pressure rose and the piston was driven outwards. On the return stroke the gases escaped through the exhaust valve. After a few revolutions the vaporiser was automatically kept hot by the heat from the burning gases. Stuart's original ideas included compression ignition, airless injection of the fuel and various forms of combustion chambers. In 1891 Hornsby and Sons, Limited, of Grantham undertook the manufacture of engines according to Stuart's patent and in the subsequent development of the "Hornsby-Akroyd" engine, Stuart took no part. The experimental work necessary was carried out by Mr W. J. Young who, on March 17, 1937, in a paper read to the Newcomen Society reviewed the many problems which had to be solved before this, the first compression-ignition oil engine, became a success.

Rudolph Diesel was six years older than Stuart, having been born in Paris of German parentage, on March 18, 1858. Sent to school at Augsburg he subsequently became a student in the Technical High School at Munich, where he attracted the attention of Carl von Linde, and it was Linde's lectures on thermodynamics which stimulated his interest in heat engines. After gaining practical experience in the shops of Sulzer Brothers, Winterthur, Switzerland, he became connected with a Paris firm engaged in the construction of Linde's refrigerating machines. On February 28, 1892, he took out his famous patent and the following year published his memoir *Theory and Construction of a Rational Heat Motor*. His ideas, based on theoretical considerations, were even more novel than Stuart's, and some of them proved impracticable. In the hands of the Maschinenfabrik Augsburg-Nürnberg, however, in the course of four years of extremely difficult experiment, the "Diesel" engine was gradually developed as a practical machine, and its debut marks an epoch in the history of power. The first successful Diesel engine, built in 1897, now stands in the Deutsches Museum, Munich. The Diesel engine differed from that of Stuart in having no vaporising chamber, while the air was compressed in the cylinder to about 500 lb. per square inch, whereby its temperature was raised to about 1000° F. and the oil was injected into the cylinder by air compressed to about 1000 lb. per square inch. Licences for the construction of Diesel engines were quickly secured by many firms in Europe and America. Diesel himself established an office in Munich, and he lived to see his engines used in every part of the world both ashore and afloat. In his own words his engine broke "the monopoly of coal". In March 1912 he came to London to read a paper on "The Diesel Oil Engine and its Industrial Importance, particularly for Great Britain" to the Institution of Mechanical Engineers. The paper was read to a crowded audience and it remains to-day of great historical importance and of not a little tragic interest. Eighteen months later he set out again to visit England, but on the passage from Antwerp to Harwich in the *Dresden* on the night of September 29–30, 1913, he disappeared from the ship and nothing more was ever heard of him.

In his paper of 1912 Diesel said "the first marine Diesel engine of 20 horse-power was constructed in 1902–3 in France, for use on a canal boat, by the French engineers Adrien Bochet and Frédéric Dyckhoff, in conjunction with the author". Adrien Bochet (1863–1922) at this time was chief engineer of the firm of Sautter-Harlé and Co., Paris, a position he held till 1908. The engine referred to was a horizontal engine with two opposed cylinders working on the four-stroke cycle. It was fitted to the barge *Petit-Pierre* belonging to the ironfounders MM. Hachette and Driout. Sautter-Harlé and Co. built other marine engines of the same type, but they soon turned to the construction of vertical engines and they were the first firm to make such engines for submarines.

Among the other pioneers of Diesel engines for propulsion were Nobel Brothers of St Petersburg, who in 1904 fitted engines of 360–450 horse-power in the oil tanker *Vandal* employed on the Volga and the Caspian Sea. These early engines were not reversible and in some cases they drove electric generators supplying current to motors for use in manœuvring and going astern. In 1905 Sulzer Brothers made a reversible two-stroke engine, and three years later Nobels made a reversible four-stroke engine. The example set by Sautter-Harlé and Co., and Nobel Brothers was soon followed by others and in an appendix to his paper of 1912, Diesel gave a list of some 300 Diesel-engined vessels of various sizes. In September 1910 the *Romagna* of 1000 tons displacement and 800 horse-power was placed in commission, and the same year the Werkspoor Company of Amsterdam built the cargo vessel *Vulcanus* of about 2000 tons displacement with a 370 horse-power six-cylinder Diesel engine. A year later the *Toiler*, built on the Tyne and fitted with a Diesel engine made by the Aktiebolaget Diesels Motoren, Stockholm, crossed the Atlantic.

The most-striking advance at this time was made by Burmeister and Wain of Copenhagen with the *Selandia*, the performance of which on a trial run from London to Antwerp in March 1912 aroused great interest among British engineers. The *Selandia*, now the *Norseman*, was one of three sister ships named after the three districts of Denmark Selandia, Fionia and Jutlandia. Built for the East Asiatic Company for trading to the Far East she is 370 ft. long, 9800 tons displacement and is driven by twin screws. Her two engines each had eight cylinders $20\frac{7}{8}$ in. diameter with a stroke of $28\frac{3}{8}$ in. and at 140 revolutions per minute developed a total horse-power of 2480, giving the ship a speed of 11 knots. Air at 300 lb. pressure for starting was supplied by

PLATE XV

DIESEL ENGINE OF SUBMARINE *EMERAUDE*, 1906
Courtesy of Sautter-Harlé and Co.

DIESEL ENGINE OF SUBMARINE *CIRCE*, 1907
Courtesy of Maschinenfabrik Augsburg-Nürnberg, A.G.

separate Diesel-driven compressors, and single-stage compressors driven off the forward end of the crankshaft provided air at 900 lb. pressure for fuel injection. The consumption of oil was 0·363 lb. per horse-power per hour. Her sister ship, the *Jutlandia*, was built on the Clyde by Barclay Curle and Co., and she was the first ocean-going motor ship constructed and engined in a British yard. The voyages of these and other vessels effectively demonstrated the suitability of Diesel engines for the propulsion of ships, and by 1914–15 Lloyd's *Register Book* contained the particulars of 297 motor ships of over 100 tons, having a total gross tonnage of 234,287 tons.

While the oil engine was thus being successfully applied to mercantile vessels, it had made but little progress in warships with the exception of submarines. It is true that a few oil-driven electric generators were to be found in capital ships, that proposals for gas and oil-driven battleships had been put forward, and that the Russian Government had fitted Diesel engines in one or two gunboats, but so far Diesel engines had not been fitted in any important warship. In submarines, on the other hand, the Diesel engine had entirely superseded the earlier petrol engines, and in the great struggle at sea of 1914–18 the Diesel engine played an important part. Just as the French were the most progressive pioneers of the submarine, so they were the first to use Diesel engines in such vessels. Through the courtesy of Sautter, Harlé and Co. and the Maschinenfabrik Augsburg-Nürnberg, photographs are given on Pl. XV of the earliest examples of four-cylinder submarine engines. Those made by Sautter-Harlé and Co. were begun in 1904 and in 1906 were fitted in the *Emeraude* and *Opale*, while those made at Augsburg were delivered in

1907 and fitted in the *Circe* and *Calypso*. The engines of the *Emeraude* and *Opale* on trial at 340 revolutions per minute developed 395 horse-power. In September 1907 the *Opale* made a successful voyage of 550 miles, and a year later the *Emeraude* a voyage of 692 miles.

The engines of the *Circe* and *Calypso*, unlike the others, were reversible. The cylinders were 13 in. diameter, the stroke about 14 in., and each engine at 400 revolutions developed 300 horse-power. In September 1909 the vessels made long voyages from Toulon. The French Minister of Marine, writing to the makers on April 11, 1910, said, "J'ai l'honneur de vous faire connaître que, depuis leur mise en service, ces moteurs ont eu un fonctionnement très satisfaisant."

The British submarines of the A, B and C classes, vessels of from 190 tons to 285 tons surface displacement and 450 to 600 horse-power, nearly all built by Vickers, Limited, during 1902–8 were driven by horizontal petrol engines. In 1903 the Admiralty had ordered a 500 horse-power four-cylinder Hornsby-Akroyd engine for trial, but it had not proved satisfactory. About this time Vickers secured a license for the construction of engines of the M.A.N. type, and a four-cylinder Diesel engine made by them was fitted in A 13 as an experiment, the work being completed in 1908. The results being satisfactory, the D class of submarines of 495 tons surface displacement were all fitted with two six-cylinder Diesel engines with a total of 1200 horse-power. Difficulty being experienced with the air compressors providing air for injecting the fuel, Vickers, Limited, evolved the so-called "common rail" system of "solid" or "airless" injection. A large amount of experimental work has been carried out on engines for submarines at the Admiralty Experimental

Station, and since the War the fast patrol submarine *Thames*, of 1805 tons surface displacement, has been fitted with engines of 10,000 horse-power, giving her a surface speed of 21¾ knots.

Before turning to record the progress of the oil engine in mercantile vessels it is perhaps desirable to point out one or two factors which rendered the construction of large oil engines much more difficult than the construction of large steam engines. In normal designs of reciprocating steam engines the highest pressure is about 250 lb. per square inch and the highest temperature about 600° F. In an oil engine the pressure rises to 550 lb. or more and the temperature to between 2500° and 3000° F., the flame of the burning gases in the cylinders being white hot and the temperature exceeding the melting-point of cast iron. The cylinders, pistons and valves of oil engines, therefore, have to be made not only strong enough to withstand very high stresses, but they must be designed so that they can be effectively cooled. These considerations place a practical limit on the size of cylinders in oil engines, which are small as compared with the steam engine cylinders shown on Table XI, p. 249. For large powers, oil engines accordingly have as many as ten or twelve cylinders. Moreover experience showed that the metals used in oil engines had to be of a special character and that the workmanship had to be of a higher order. Though at first most oil engines were constructed to work on the single-acting four-stroke principle, so familiar through the widespread use of four-stroke petrol engines in motor cars, engines are made working on the double-acting four-stroke cycle and on the single-acting and double-acting two-stroke cycle. Each of these designs brought its own special problems the solution of which has led to the construction of engines of many types. With the development of these various types has come the practice of forced "scavenging" by which the cylinders are effectively cleared of the burnt gases, and also the practice of "supercharging" or "pressure-charging" by which a greater weight of air is forced into the cylinder before compression begins. This enables more fuel to be burned and the power to be increased. "Scavenging" and "supercharging" are carried out in a variety of ways, one of the most interesting systems of supercharging being that of Mr A. Buchi, in which the exhaust gases from the engine are made to drive a gas turbine, which itself drives a turbo-blower delivering air at a small pressure to the cylinders.

The total number and tonnage of ships with oil engines included in Lloyd's *Register Book* for 1935–6 was 6128 ships and 12,290,599 gross tons. In these ships are found about 20 types of engines but according to Lloyd's *Shipping Index* for March 9, 1937, taking all steel vessels of over 2000 tons in actual service, 718 had Burmeister and Wain engines, 241 Sulzer engines, 201 M.A.N. engines, 140 Doxford engines, and 101 had Werkspoor engines.

Burmeister and Wain acquired the rights of the Diesel patent in 1898, five years later produced their first really reliable engine for land purposes and in 1911 built the marine engines for *Selandia* as mentioned above. These engines were of the four-stroke type, but engines working on the two-stroke cycle have been developed and airless injection and supercharging adopted. Licences for constructing Burmeister and Wain engines have been granted to many firms. A notable early post-war vessel built on the Tyne but engined by Burmeister and Wain was the *Gripsholm* of 17,716 gross tons. She was fitted with two

PLATE XVI

12,000 HORSE-POWER TEN-CYLINDER OIL ENGINE
FOR *STIRLING CASTLE*, 1936

Courtesy of Harland and Woolf, Ltd.

double-acting four-stroke six-cylinder engines developing a total of 13,500 horse-power giving the ship a speed of 17 knots. The first ship with a Burmeister and Wain double-acting two-stroke engine was the single-screw vessel *Amerika* of 10,000 tons and 7000 horse-power built in 1929. The principal builders of Burmeister and Wain engines in Great Britain are Harland and Wolff, Ltd., of Belfast. In 1930–2 this firm built and engined the large Cunard White Star liners, *Britannic* and *Georgic*, of 27,000 gross tons. Each of these vessels had two double-acting four-stroke ten-cylinder engines with a total horse-power of 20,000. More recently the same firm have built the Union Castle liners, *Stirling Castle* and *Athlone Castle*, 725 ft. long, 25,550 gross tons, driven by double-acting two-stroke engines developing 24,000 horse-power. A photograph of one of the engines of the *Stirling Castle* is shown on Pl. XVI. The engine has ten cylinders, 26 in. diameter by 4 ft. 11 in. stroke. It is 34 ft. high from the centre of the crankshaft, its length including the thrust block is 72 ft. and its weight 900 tons. So far this is the largest marine oil engine constructed in Great Britain. As an example of the method of construction of a piston for such an engine a sketch is given in Fig. 45. The top and bottom portions of the piston are of heat-resisting

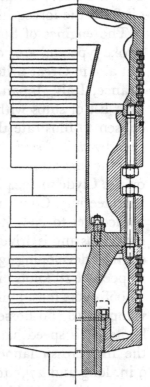

Fig. 45. Oil engine piston

chromium steel, while the centre portion is of special cast iron. The grooves for the piston rings are formed by landing rings of perlitic iron caulked into grooves in the chromium steel portions. The steel piston rod has an outer sleeve of cast iron, the whole being cooled by oil from the forced lubrication system.

The engines of Sulzer Brothers are all of the two-stroke type and are constructed not only at Winterthur but also in Great Britain, Holland, Belgium, Germany, France, Italy, Japan and the United States. From a long list of ships with Sulzer engines three have been chosen to illustrate the development with this type— the *Aorangi*, *Victoria* and *Prince Baudouin*. The *Aorangi* was the first large motor liner to be built. Constructed on the Clyde in 1924 by the Fairfield Shipbuilding and Engineering Company, Limited, for the run from Vancouver to New Zealand, she is 600 ft. long, 17,491 gross tons and is driven by four single-acting two-stroke engines developing 13,000 horse-power, giving the ship a speed of 18 knots. In 1931 the *Victoria*, 534 ft. long, 13,068 gross tons, built at Trieste with four engines developing 18,660 horse-power attained a speed of 22·86 knots. This speed has since been surpassed by that of the fine cross-channel vessel *Prince Baudouin*, 370 ft. 9 in. long and 2755 tons displacement, fitted with two twelve-cylinder single-acting two-stroke engines, which at 268 revolutions per minute develop 17,000 horse-power, giving a maximum speed of $25\frac{1}{4}$ knots. The service speed is 22 knots. The cylinders of the engines are 22·8 in. diameter, 33·1 in. stroke, the piston speed being 1480 ft. per minute.

As the pioneers of the Diesel engine, the Maschinenfabrik Augsburg-Nürnberg has a long record of experimental work to its credit, and since the War it has been

the pioneer in the use of high-speed oil engines in combination with reduction gearing for ship propulsion.
During the War the M.A.N. and its licensees constructed submarine engines of more than half a million
total horse-power. In 1919 many of these engines became available for commercial purposes and some were
fitted in merchant ships in combination with gearing.
Early examples of geared Diesel engines were seen in
the Hamburg-American liners *Havelland* and *Munster-
land*, 6300 gross tons, which in 1921 were each fitted
with two ten-cylinder 1500 horse-power engines driving
the propeller shafts through reduction gearing having
a ratio of about $2\frac{1}{2}$ to 1. In 1924 the *Monte Olivia* and
*Monte Sarmiento*, 13,600 tons, were each fitted with four
engines geared in pairs to the propeller shafts, and in
1929 the *St Louis* and *Milwaukee*, 16,700 tons, had
machinery of the same type of 12,600 horse-power
giving the ships a speed of 16 knots. It was the experience gained with these installations that enabled
the M.A.N. to solve the problem of driving a battleship
with oil engines, the *Deutschland* and *Admiral Graf Spee*
both having Diesel engines and gearing. The total
horse-power in these vessels is 56,000. Each ship has
eight nine-cylinder double-acting two-stroke engines of
7100 horse-power, coupled in groups of four to two propeller shafts. The engines run at 450 revolutions per
minute and the propellers at 250 revolutions. Electricity for the ships is generated by eight 400 horse-
power Diesel engines, running at 1000 revolutions per
minute, the whole being as *The Motor Ship* said "a
remarkable and entirely novel production". Only experience can show whether such a plant is equal in all
respects to a steam turbine plant. Of the larger ships
fitted with direct driving M.A.N. engines, mention

should be made of the Italian liner *Augustus* of 32,500 tons and 28,000 horse-power built in 1927, which at present is the largest motor vessel afloat.

Developed somewhat later than the engines already dealt with, the Doxford oil engine, since it was introduced in 1921, has been fitted in 154 ships having an aggregate of 584,380 horse-power. The design of this engine is due to Mr C. Keller of William Doxford and Sons, Limited, and in many respects it is unique. The engine is of the opposed piston type and works on the two-stroke cycle, an interesting feature being that the upper pistons have only about three-quarters of the stroke of the lower pistons. Engines are made with three, four, five or six cylinders and the balancing is so good that it is possible to run an engine on the test bed without holding-down bolts. The first vessel with a Doxford engine was the *Yngaren*, 9000 deadweight tonnage, which had a four-cylinder engine which at 78 revolutions developed 3000 horse-power. In 1927 the passenger liner *Bermuda* was fitted with four engines having a total of 13,500 horse power. The largest Doxford engines built so far are those in the *Essex* and *Sussex* constructed in 1936 by John Brown and Company, Limited, for the New Zealand Shipping Company, for the transport of frozen cargoes from Australia and New Zealand. These vessels are 530 ft. long and when fully loaded have a displacement of 24,668 tons. They are each driven by two five-cylinder engines which at their maximum develop a total of 14,400 horse-power. The cylinders are about $28\frac{1}{2}$ in. diameter, the upper pistons have a stroke of about $37\frac{1}{2}$ in. and the lower a stroke of about 51 in. The speed of the *Essex* on trial was 19·5 knots; the service speed being 17 knots. As in many other motor ships, the waste gases from the engines are used in boilers for generating steam for subsidiary purposes.

As with the steam turbine, the high-speed Diesel engine running at 1000 to 2000 revolutions a minute can be employed for propulsion by the use of electric transmission gear; and this system has been adopted in a few vessels of moderate size and in tugs. One aspect of the possible development with Diesel electric plant was dealt with by Mr H. C. Ricardo in his lecture on "High-Speed Diesel Engines for Marine Purposes", delivered to the Institution of Mechanical Engineers on December 1, 1933. He found it difficult, he said, to make out a case for a Diesel engine of any sort in really large or very fast vessels, and the monster Diesel engine in which hundreds of tons of metal were lavished to secure rigidity he regarded as an anachronism like the prehistoric monsters in the Natural History Museum. For a vessel requiring about 6000 horse-power for all purposes, however, he suggested the installation of some seventy-five 100 horse-power high-speed Diesel-driven generators delivering current to two propulsion motors and auxiliary plant. The units would be identical and interchangeable, each in its own sound-proof container and easily removable from the ship. Reserve sets would be kept in a central depot where necessary overhauls would be carried out.

CHAPTER XXII

# MARINE ENGINEERING AND
# THE NATION

THOUGH any history of marine engineering must necessarily be devoted mainly to the invention, introduction and development of machinery, there are other matters connected with the subject which should not go unnoticed. Among these are the influence on progress of government departments, classification societies, and technical institutions, the benefits derived from scientific investigations and the inter-relation between the shipping, shipbuilding and marine engineering industries.

The government department which has most profoundly influenced the progress of marine engineering is undoubtedly that of the Admiralty. The employment of steam in naval vessels more than a century ago led to the engagement of men familiar not only with the working of machinery, but with its design, construction and repair. As steam extended its domain so the Admiralty in time became the largest, as it is to-day the oldest, steamship owner in the world. But the Admiralty was interested in engineering before the days of steam navigation, and the first engineer to be employed by the old Navy Board was Sir Samuel Bentham (1757–1831). In 1799, as Inspector-General of Naval Works, Bentham installed a 12 horse-power pumping engine for the docks at Portsmouth, and afterwards erected steam-driven saw mills and metal mills. To him is due some of the credit for the success of the famous block-making machinery at Portsmouth, devised by the elder

Brunel and made by Henry Maudslay. In 1796 Bentham engaged as his assistant Simon Goodrich (1773–1847), who in 1814 became Mechanist to the Navy Board with his headquarters at Portsmouth, and he held this post till 1831. Only those who have read his note-books, preserved in the Science Museum Library, can fully appreciate the range of Goodrich's activities. He was the trusted adviser of the Navy Board on all engineering matters, including the machinery of ships. Four years after Goodrich retired, the Admiralty in 1835 appointed Peter Ewart (1767–1842) Chief Engineer and Inspector of Machinery at Woolwich Dockyard, the first naval yard to have shops for the repair of boilers and engines. Ewart had as his assistant Thomas Lloyd (1803–75) who after being trained as a shipwright at Portsmouth, had been detailed to study steam engines and machinery, and who ultimately became the first Engineer-in-Chief of the Navy. In 1837 when the Admiralty took over the Post Office Steam Packets, they appointed Captain (afterwards Admiral) Sir William Edward Parry (1790–1855) Superintendent of Steam Machinery and Packet Service at Whitehall. Parry, retiring in 1846, was succeeded by Captain Alexander Ellice who, however, held office but a short time, and in 1847 Lloyd was transferred to Whitehall from Woolwich with the title of Chief Engineer and Inspector of Machinery, a title afterwards changed to that of Engineer-in-Chief of the Navy. As a youth at the School of Naval Architecture at Portsmouth, Lloyd had seen the birth of the steam navy, and when he retired in 1869 the steam-driven mastless iron-hulled iron-armoured vessel *Devastation* was about to be laid down at Portsmouth. He had been responsible for the introduction of tubular boilers and superheaters, the screw propeller, compound engines and surface condensers. He had had control of the largest body of civilian marine engineers in the country, and in his time the naval engineering branch had grown into a force of about 1000 officers and 4000 men. "During a very long public life", wrote the Controller of the Navy, "Mr Lloyd has been distinguished for wise and carefully considered suggestions for the improvement of the details of marine engines, and I venture to say that the principal marine engine makers in the kingdom have frequently consulted him, and always benefited from his advice."

Lloyd had been head of the "Steam Branch" at the Admiralty directly under the Controller, but after his retirement the rank of Engineer-in-Chief was abolished, a "Constructive and Engineering Staff" was formed and in this his successor James Wright (1824?–99) was only granted the rank of Engineer Assistant. In 1872 the title of Engineer-in-Chief was restored and this Wright continued to enjoy till his retirement in 1887. He had joined the staff at Woolwich Dockyard in 1845 and for many years had been Lloyd's assistant at headquarters. It was during his period of office that the triple-expansion engine and the torpedo boat were introduced. On Wright's retirement the post of Engineer-in-Chief passed to a naval engineer officer, Inspector of Machinery Richard Sennett (1847–91), who had been a student at the Royal School of Naval Architecture and Marine Engineering and had a brilliant career in the service. Sennett held office only two years and was succeeded in 1889 by Inspector of Machinery (afterwards Engineer Vice-Admiral Sir) John Durston (1846–1917). Taking office when great demands were being made for increased speed in ships,

Admiral Durston was responsible for the introduction of the water-tube boiler and the steam turbine in all classes of vessels. His immediate successors were Engineer Vice-Admirals Sir Henry Oram and Sir George Goodwin, who held office from 1907 to 1917 and from 1917 to 1922 respectively. On them fell the onerous duty of maintaining the efficiency of the engine-room department of the Royal Navy during the Great War, when the number of vessels on the Navy List rose to over 1000 and the engineering personnel on active service to nearly 5000 officers and 86,000 men. Since Admiral Goodwin's retirement in 1922 the office of Engineer-in-Chief of the Fleet has been held in turn by Engineer Vice-Admirals Sir Robert Dixon, Sir Reginald Skelton, and Sir Harold Brown, the present Engineer-in-Chief being Engineer Vice-Admiral G. Preece, who was appointed in 1936.

Anyone at all acquainted with the history of marine engineering is fully aware of the important influence exerted by the Admiralty on the design of boilers and engines and all that appertains to them, and of the large amount of experimental work carried out. Hedged around by restrictions, which do not apply to the machinery of mercantile vessels, the designer of machinery for warships is continually being pressed to reduce the weight and increase the power of the machinery, and naval engineering practice has thus reacted on marine engineering practice in many ways. Through the engineering staffs at Whitehall and the Dockyards, and the engineer overseers stationed in various districts, the Admiralty is brought into the closest touch with every development and the standards it sets are the highest attainable. Writing in 1893 when great publicity was given to all defects in the machinery of Her Majesty's ships "be they ever so small or un-important", Staff-Engineer (afterwards Engineer Rear-Admiral) W. L. Wishart (1853–1914) said that "page after page could be written in which it could be shown how Admiralty practice is not only keeping well ahead of the times, but that it influences in every way the marine engineering practice of the country". The same remark can be made to-day.

The other Government department concerned with marine engineering is the Board of Trade, with the rules of which every British mercantile vessel must conform. The construction, equipment and navigation of merchant ships has always been a matter for private enterprise, but the regulation of shipping and the safety of ships have from very early times given rise to state action in the interests of the community, and the authority exercised by the Board of Trade is but the continuance of the usage of centuries. The Rhodian Sea Law dates from about 800 B.C. and this has formed the foundation of all subsequent laws. In Roman times and in the age of the Italian republics there were rules as to measurement, seaworthiness, masts, cables, crews, passengers, collisions, wrecks, salvage and the like. The first instance of state action regarding steam vessels and their machinery in Great Britain arose out of the boiler explosion in 1817 aboard the Norwich and Yarmouth steam packet. This vessel had a cylindrical boiler 8 ft. long and 4 ft. diameter with one end of cast iron. The machinery had been made by Matthew Murray (1765–1826) of Leeds. On the morning of April 7, 1817, just after the vessel had left the landing place at Norwich, the boiler exploded killing eight persons and injuring many others. A Select Committee of the House of Commons was appointed to enquire into the accident.

After investigation, the Committee recommended that all steam packets should be registered; that boilers should be of wrought iron or copper; that every boiler should be inspected by a skilled person; that every boiler should have two safety valves, one of which should be inaccessible to the engine man, and that an inspector should examine the valves and determine for what pressure they should be set. A penalty was to be imposed for increasing the weight on the safety valve. At this time there were several departments dealing with ships, and years passed before these reasonable recommendations were adopted. Various Shipping Acts were passed and there were enquiries into accidents, but for the time the suggestions bore little fruit. In 1846 an Act was passed for "regulating the construction of Sea Going Steam Vessels and for preventing the occurrence of Accidents in Steam Navigation". From this sprang the extensive system of Government supervision now in operation. Among other things, the Act laid down that passenger vessels were to be surveyed half-yearly by surveyors approved by the Board of Trade. Another important Act was the Mercantile Marine Act of 1850 by which the Marine Department, now the Mercantile Marine Department, of the Board of Trade was established to take over the duties hitherto performed by at least nine different authorities. In 1851 the Steam Navigation Act was passed by which the Board of Trade was empowered to appoint, instead of merely approving, surveyors. It also enacted that after March 31, 1852, it would not be lawful for any steam boat of which a survey was required to go to sea or steam upon a river "without having a safety valve upon each boiler, free from the care of the Engineer, and out of his control and interference". On December

22, 1851, Robert Galloway wrote from Surrey Boat House, Lambeth, accepting the office of surveyor for London. Another of the early surveyors was Robert Murray, the author of the *Rudimentary Treatise on Marine Engines*, referred to on p. 125. In 1854 a Merchant Shipping Act was passed by which all previous shipping acts were consolidated and extended in a code of 548 clauses. This was the principal Act until another bearing the same name was passed in 1894. Gradually during the period 1854 to 1894 a code of rules was evolved dealing with nearly all kinds of marine machinery. Foremost among them were the rules relating to boilers, valves, gauges, pumps and shafting. Surveyors were empowered to attend steam trials and issue certificates of seaworthiness. Examinations for sea-going engineers were introduced in 1863. The first rules relating to boilers were issued in the early seventies when William Thomas Traill (1829–1910) was the Chief Surveyor. Traill had entered the Navy in 1853 and between 1853 and 1864 had served in the *Odin, Retribution, Edinburgh* and *Jackal*. Promoted to Chief Engineer in April 1864 he accepted the position of Engineer Surveyor and Examiner of Engineers to the Board of Trade. In 1870 he was placed on the retired list of the Royal Navy, on which he was advanced to the rank of Fleet Engineer in 1886. Besides his post at the Board of Trade he also held that of General Superintendent of Proving Machines for Lloyd's Register of Shipping. On October 21, 1874, the same day as William Parker the first Chief Engineer Surveyor of Lloyd's Register, he gave evidence before the Admiralty Boiler Committee, and he presented to the Committee a copy of the Boiler Rules of the Board of Trade. In 1888 he published his *Boilers, Marine and*

*Land.* Previous to this he had, in 1885, published his standard work on *Chain Cables and Chains.*

Traill retired in 1896, having been Engineer Surveyor-in-Chief since April 1, 1876, and was succeeded by Peter Samson (1842–1909) who had also served in the Navy but had joined the Staff of the Board of Trade in 1873 and for many years had been Traill's assistant. Samson was a member of the Admiralty Committee on the design of engines and boilers appointed in 1892 and presided over by Admiral Buller. Samson's successors have been Alex Boyle (1850–1929), Thomas Carlton (1861–1937), Albert Edward Laslett (1866–1937), Mr W. McAuslan and the present Engineer Surveyor-in-Chief, Mr W. T. Williams. In 1936 the Engineering Staff of the Board of Trade included beside the Engineer Surveyor-in-Chief and his deputy, nineteen Senior Engineer Surveyors and fifty-nine Engineer Surveyors. The regulations which guide them in their duties are set out in the *Instructions as to the Survey of Passenger Steamships* published by His Majesty's Stationery Office.

The third important organisation in Great Britain which exerts a marked influence on the advancement of marine engineering is Lloyd's Register of Shipping, the most famous of ship classification societies. The practice of classifying ships according to their condition arose out of the needs of shipowners and underwriters and the age-long practice of insurance. Marine insurance is probably as old as maritime law, but the great insurance corporation of Lloyd's with its 1400 underwriting members and its 1500 agents at home and abroad, and Lloyd's Register of Shipping with its 400 surveyors—entirely independent institutions—both had their birth in the seventeenth-century London coffee house kept by Edward Lloyd, and it is from him their names are derived. How the histories of these important institutions are interwoven can be seen by reading *The Romance of Lloyd's* by Commander F. Worsley and Captain G. Griffith, and the *Annals of Lloyd's Register*, the centenary number of which was published in 1934.

From the latter it will be seen that the compilation of lists of ships was begun by frequenters of the coffee house who were interested in shipping. When the first lists were made is not known, but the earliest in existence bear the dates of 1764–65–66. Lloyd's Register is in direct descent from the body of men who formed these registers, although the society as known now was established on its present basis only in 1834. Although Lloyd's Register is the oldest and most influential classification society, it has no monopoly, and nearly every important maritime country has its own society, while in 1890 at Glasgow was founded the British Corporation of Shipping with similar aims. The societies abroad include the Bureau Veritas founded in 1828, the Registro Italiano, Navale ed Aeronautico, the American Bureau of Shipping, Det Norske Veritas, and the Germanischer Lloyd, founded in 1861, 1862, 1864 and 1867 respectively. Classification by one or other of these societies indicates that a vessel has been built and maintained in accordance with its rules.

When Lloyd's Register was reconstituted in 1834 under the name of Lloyd's Register of British and Foreign Shipping, a name it retained until 1914, the Society had a proper organisation for the survey of hulls and fittings of ships, but not of their machinery, and the first rules relating to steamships contained the simple provision:

All seagoing vessels navigated by *Steam* shall be required to be surveyed *twice in each year*, when a character shall be assigned to them according to the report of survey as regards the classification of the hull and materials of the vessel.

With respect to the Boilers and Machinery the Owners are required to produce to the Surveyors to this Society at the above-directed surveys, a certificate from some competent *Master Engineer*, describing their state and condition at those periods.

When the last clause had been complied with, a notation of M.C. (Machinery Certified) was assigned against the vessel's name in the *Register Book*.

This plan of relying on survey by independent surveyors survived for forty years. In 1865 a proposal made to the Society, that it was desirable that marine machinery should be surveyed by the Society's Surveyor during construction was discountenanced, but in 1869 it was stipulated that in future the engines and boilers would be considered as part of the vessel's equipment, and that figure " 1 " would be likely to be withheld if the condition of the machinery was found to be unsatisfactory. But the number of steamships and the size of machinery were fast increasing and in 1874, no doubt influenced to some extent by the activities of the Board of Trade, the Managing Committee decided to appoint a fully qualified Engineer Surveyor to promote the organisation of the survey of machinery. Their choice fell on William Parker who had been a Board of Trade Surveyor. At the same time Surveyors were appointed at Liverpool and on the Clyde. In 1875 the staff was further increased by the appointment as assistant to Parker in London of the young naval engineer James Taylor Milton (1850–1933) who was destined to increase greatly the prestige of the engineering branch of the Society. Born at Portsmouth and trained as an engineer student, Milton had gained a scholarship to the Royal School of Naval Architecture and Marine Engineering, the work of which, as before stated, was transferred in 1873 to the Royal Naval College, Greenwich. As an assistant engineer Milton left the College with the highest certificates and in October 1874 was appointed to the Indian troopship *Jumna* where he had as his shipmates Alfred Morcom, R. J. Butler and J. T. Corner. In May 1875 he left the naval service and from that time onward till 1921 except for four years, 1886–90, during which he was manager for R. and W. Hawthorn, Leslie and Co., Newcastle-upon-Tyne, Milton was in the service of the Society, succeeding Parker as Chief Engineer Surveyor in 1890. He assisted Parker to draw up the rules for boilers issued by the Society in 1876, and the following year read his first paper to the Institution of Naval Architects. In 1878 Parker likewise contributed a paper to the same Institution, and these are the first of a long series of contributions to engineering literature by the staff of the Society. In 1879 regulations were issued for the periodical inspection of the machinery of ships, and by 1884 one-third of the Society's technical staff were engineers. Rules for shafting were issued in 1884, for electrical installations in 1891, for refrigerating machinery in 1898, for petrol and paraffin engines in 1910 and for Diesel engines in 1914. Under Milton and his colleague Sir Westcott Abell, the Chief Ship Surveyor, during the War the Society's staff did work of the greatest importance to Great Britain and her Allies. Milton retired in 1921 and he was succeeded by Harry Alfred Ruck-Keene (1865–1932) who had joined the staff in 1890, and who at his death was succeeded by the present Chief Engineer Surveyor Dr S. F. Dorey.

Writing in 1926, Sir Andrew Scott, long the Secretary of Lloyd's Register, said that there are three conditions essential to the success of a classification society: first, the governing body must adequately represent the several interests concerned; secondly, its rules must be based upon the best theoretical knowledge and practical experience, and must be kept up-to-date; and thirdly the staff must be composed of men competent to interpret and apply the rules with intelligence and impartiality. It is the compliance with these conditions which has enabled Lloyd's Register of Shipping to establish a standard of merchant ship-building and marine engineering accepted throughout the world.

Those who have read the preceding chapters will have observed that over and over again reference has been made to the *Transactions* of technical societies. Without these transactions it would indeed be impossible to write a history of any branch of engineering. It is therefore only fit that a few notes should be given regarding those Institutions whose work has been closely associated with ships and their machinery.

The earliest of such societies was that founded in London in 1791 and called the Society for the Improvement of Naval Architecture. It had some 300 members and the Duke of Clarence was its President. Though it only survived eight years it left its mark behind, for it was responsible for initiating the experiments of Colonel Henry Beaufoy on the resistance of bodies moving through water, the results of which were used by Fulton in determining the power required to drive steam boats. To its influence can be traced the birth of the first School of Naval Architecture at Portsmouth. There had been an earlier engineering society, called the Society of Civil Engineers, but the members of this were concerned with roads, bridges, harbours and docks, and not with ships. It is with the Institution of Civil Engineers that the history of the modern technical society begins. From the first the Institution included among its members marine engineers, and its earlier *Proceedings* contain information relating to their work. The most remarkable Presidential address ever given to an engineering society is probably that of Sir William White who, in 1903, dealt with the advancement of naval architecture, shipbuilding and naval and marine engineering during his lifetime. This address occupied 170 pages of the *Proceedings*. The second technical society of importance to be founded was the Institution of Mechanical Engineers which had its birth at Birmingham in 1847. George Stephenson was the first President, and his son Robert the second. Its fifth President was John Penn and its eighth Robert Napier. The earliest marine engineering paper read to the Institution was that by Andrew Lamb who, in 1852, described the boiler he had introduced into the ships of the Peninsular and Oriental Steam Navigation Co. The *Proceedings* for 1856 and 1858 contain Penn's account of the invention of the lignum-vitae bush; those of 1867 Macfarlane Gray's account of the invention of the steam steering engine. In more recent times the Institution has carried out, in co-operation with others, comparative trials of steam ships and motor ships and has stimulated research in many directions. While the Institution of Mechanical Engineers was founded just when the country was being covered with railways, the Institution of Naval Architects came into being when for ships, steam was superseding sail, and iron supplanting wood. "It was", said Sir William White, "founded at a psychological moment and all classes interested in shipbuilding welcomed it." The inaugural meeting was held on June 16, 1860. Eighteen persons were present, including ten who had been or were still employed as constructors by the Admiralty, and among the others were Scott Russell, John Penn, John White, of Cowes, and John Grantham. The meeting passed the resolution: "We who are present do now constitute ourselves an Institution of Naval Architects for the purpose of advancing the science and art of naval architecture". The first President was Sir John Pakington (afterwards Lord Hampton) who, as First Lord of the Admiralty, had been responsible for ordering the construction of the *Warrior*. He was succeeded by Lord Ravensworth, and since then the Presidency has always been held by a nobleman interested in shipping. Though from the first, papers on marine engineering subjects were read, it was not till 1869 that marine engineers were admitted to full membership. Since then marine engineers have formed a considerable proportion of the membership, and there are few important advances in marine engineering not recorded in the *Transactions* of the Institution.

Five years after the formation of the Institution of Naval Architects in London, the Institution of Engineers and Shipbuilders in Scotland was founded at Glasgow by the amalgamation of the Institution of Engineers in Scotland and the Scottish Shipbuilders' Association. The former of these had been inaugurated in 1857 with Rankine as its President, and the latter in 1860 with Robert Barclay as President. The President of the new society was J. G. Lawrie (d. 1891). The example set by the engineers and shipbuilders on the Clyde was followed in 1884 by those of the Tyne and the Wear, when at Newcastle-on-Tyne was formed the

North East Coast Institution of Engineers and Ship-builders with William Boyd (1839–1919), of the Wallsend Slipway and Engineering Co., as its President. The promoters of these various Societies were chiefly concerned with the design and construction of ships and machinery, and not with their operation, but in 1889 a group of superintendent engineers of steamship companies in London met at Stratford and founded the Institute of Marine Engineers which to-day, through its practice of issuing its *Transactions* in monthly parts, enables its members in ships all over the world to be kept informed of the latest developments. The Institute has had many eminent men among its Presidents and members, but to no one has the Institute been more indebted than to James Adamson (1850–1931), who for forty years was its Honorary Secretary. There are many other bodies both at home and abroad which in one way or another further the practice of marine engineering, among the most prominent of those in foreign countries being the Association Technique Maritime founded in Paris in 1888, the American Society of Naval Architects and Marine Engineers incorporated in 1893 and the Schiffbautechnische Gesellschaft founded in Berlin in 1899.

In no direction have technical societies done more to further the progress of marine engineering than by promoting the application of scientific research. To-day all the weapons in the armouries of mathematicians, chemists, physicists and metallurgists are used in the attack of practical problems. Scientific research has spread from private laboratories to the laboratories of Government departments, industrial research associations and firms, and it is in the *Proceedings* of the technical institutions that the results of the investigations are first published. Corrosion, fuels, combustion, lubrication, heat transmission, the properties of steam, the flow of liquids and gases, the constitution of metals and alloys, the behaviour of materials under high pressures and temperature, and the oscillation and vibration of machinery are but a few of the subjects, the study of which has enabled the designer and constructor to increase the efficiency of boilers, steam engines and oil engines. But the efficiency of sea transport depends also on other things than the efficiency of the machinery. The engineer may reduce the size of his machinery and the amount of fuel it consumes, and so enable larger cargoes to be carried, he may reduce its first cost, and its cost of maintenance, but much of his effort may be wasted if the ship is not properly designed. Here again scientific methods have been adopted with far-reaching results and the ships of to-day can be driven with far less power than they could be a century ago. This improvement is largely due to experiments made with model ships in what are known as experimental tanks.

Among the earliest recorded ship model experiments are those of Benjamin Franklin* who, after a trip to Holland where he had noticed the reduction in the speed of a canal boat when the water was low, made a trough of wood 14 ft. long, 6 in. wide and 6 in. deep fitted with a loose board by which the depth of water could be varied. In the trough he placed a model boat 6 in. long to which was attached a silk cord which ran over a brass pulley fixed at one end of the trough and which was weighted with a shilling. Franklin found that the speed of the boat in $1\frac{1}{2}$ in. of water was about one-quarter less than when the water was $4\frac{1}{2}$ in. deep.

* *Nature*, July 1, 1922, p. 10.

Probably the first experiments with a model of a steam boat were those made by David Napier in Camlachie Burn when he was building the *Rob Roy* in 1818.

The modern experimental tank, however, is due to the genius of William Froude (1810–79)* who, in 1856, it is said, was asked by Brunel to study problems of stability and resistance in connection with the *Great Eastern*. His first model experiments were made in a large storage tank at the top of his house at Paignton, Devon, where he had gone to live in 1859. In 1867 he moved to a new house called "Chelston Cross" at Cockington, Torquay, and it was on land close to this he was able to build his first experimental tank. The expense of the tank was borne by the Admiralty. The tank was 278 ft. long, 36 ft. wide, and the depth of water 10 ft. Above the water was supported a track along which the towing and recording apparatus ran. The ship models were 12 ft. long and were made of paraffin wax. Many Governments and private firms now have experimental tanks far larger and better equipped than Froude's was, but in all of them the methods used are those he introduced. Investigations are made on the rolling and resistance of ships in smooth and rough water, the steering and manœuvring of ships, the resistance of bilge keels and the bossings for twin screws, on towing and on screw propellers and paddle wheels. During 1935 at the Froude Laboratory of the National Physical Laboratory, Teddington— which owes its existence to the generosity of Sir Alfred Yarrow—no fewer than seventy-three ship designs were submitted for test and modification, involving the testing of 160 model hulls. Tests are carried out on every

* See Sir W. S. Abell's Presidential Address to the Devonshire Association, July 4, 1933.

type of craft from liners and battleships to trawlers, tugs and barges. In connection with the design of the *Queen Mary*, John Brown and Co., in their own experimental tank, made 8000 experiments with model hulls, and nearly 1000 experiments with screw propellers. In a leading article on Froude, *The Engineer*, on July 13, 1934, said: "We suppose no useful estimate of the economy which has resulted from tank testing can ever be made, but it may at any rate be said that it would have to be made in millions of pounds sterling, no lesser unit being adequate for the magnitude of the computation."

Finally, a word or two may be said about the interrelation of the shipping, shipbuilding and marine engineering industries of the world and the position Great Britain holds in them. Shipbuilding and marine engineering are both handmaids of shipping and commerce, and their prosperity is entirely dependent on economic conditions. In these industries there have always been periods of great activity and of comparative stagnation, and it may be presumed there will be again. But the abnormal conditions due to the Great War, the development of engineering abroad, the rise of economic nationalism and the lavish grant of subsidies in some countries in the interests of shipping, have brought about a permanent change in the relative position of maritime countries as regards the industries referred to. In the chart shown in Fig. 46, prepared from Lloyd's Register statistics, is shown the total amount of tonnage of all vessels over 100 tons launched in each year from 1901 to 1936, which illustrates the great fluctuation in output. The chart may be supplemented by some figures. In 1894, out of a world tonnage of 1,323,538 tons launched, Great Britain and

Ireland's share was 1,046,508 tons, Germany's 117,702 tons, that of the United States 66,894 tons, and Japan's 3173 tons. In 1919, when efforts were being made to replace the ships lost in the War, the tonnage launched was 7,144,549 tons, out of which Great Britain and Ireland were responsible for 1,620,442 tons, the United

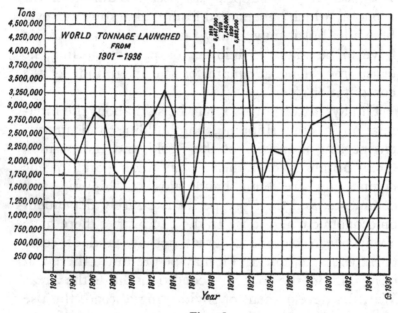

Fig. 46

States 4,074,385 tons, and Japan 611,883 tons. The total tonnage then began to assume more normal proportions until 1931 when it dropped sharply, and in 1933 the total tonnage was only 489,016 tons, or less than one-fourteenth of what it had been in 1919. With the revival of trade, there was a gradual improvement all round and the world's tonnage launched rose in 1936 to 2,117,924 tons. The contributions to this of the leading shipbuilding countries were:

| Great Britain | ... | ... | 856,257 tons |
| Germany | ... | ... | 379,981 „ |
| Japan | ... | ... | 294,861 „ |
| Sweden | ... | ... | 154,044 „ |
| United States | ... | ... | 111,885 „ |
| Denmark | ... | ... | 97,537 „ |
| Holland | ... | ... | 93,831 „ |
| France | ... | ... | 39,208 „ |

These figures show the international character of the shipbuilding and marine engineering industries and are in marked contrast to those for fifty years ago, when Great Britain built about two-thirds of the world's tonnage. There may be some who regret that Great Britain does not occupy the same outstanding position as she did last century, but it is not surprising that other countries are as desirous as we are of building their own ships, or that they can rival us in the construction of machinery. As in the case of the steam engine, the locomotive and the railway, so in the case of the ocean-going steamship Great Britain was the pioneer and it was to this country that other builders came for instruction. It is no uncommon thing for a pupil to attain to an equal or even greater fame than his master, but his debt to his master remains. In the matter of ship building and marine engineering the whole world is Great Britain's debtor. When Germany was striving to gain a place among industrial nations it was to England her engineers came to learn our methods. Carl Ziese, the collaborator of Schichau, at Elbing, and the first in Germany to construct a marine triple-expansion engine, was one of the last pupils of John Elder; while Robert Zimmermann, the constructor of the *Kaiser Wilhelm der Grosse*, the *Deutschland* and other great ships of which Germany was so justly proud, had spent eleven years at Greenock, Jarrow and Barrow, and at Barrow had been assistant to William John (1845–90), the designer of the beautiful *City of Rome*. When the Italians wanted to be able to construct marine engines for their battleships it was to Maudslays they went for their technical staff; and it was Charles Brown (1827–1905) who, after learning his profession under Charles Sells at Lambeth, gained a European reputation for the Swiss firm of Sulzer Brothers. Henry Sulzer-Steiner said of Brown that he was "no less than the founder of mechanical industry in Switzerland". Brown's two sons, C. E. L. Brown and Sydney Brown, were the founders of Brown, Boveri and Company of Baden. No shipbuilding country owes more to Great Britain than Japan, to which, in the eighteen-seventies went Ewing, Perry, Ayrton, Dyer, Milne and other men afterwards eminent in scientific and engineering circles. It was they who inaugurated the work of the Tokyo Imperial University, where no name is held in higher honour than that of Charles Dickenson West (1847–1908) who for twenty-five years taught mechanical and marine engineering, and whose bust now stands in the University grounds. The eminence attained by the shipbuilders and marine engineers of Japan is a testimony alike to their own qualities and to those of their instructors. In the case of individuals the master grows old and has to give way to others, but in the case of nations the stream of life is constantly being renewed, and in spite of what is now being done abroad, we cannot doubt that the future shipbuilders and marine engineers of Great Britain will uphold the high traditions of that great branch of engineering of which it has been the object of these chapters to trace the history.

# INDEX

Printed in the United States
By Bookmasters